Sixth Edition

CHEMICAL PRINCIPLES IN THE LABORATORY

Some material in this work previously appeared in CHEMICAL PRINCIPLES IN THE LABORATORY, Fifth Edition, copyright © 1989, 1985, 1981, 1977, 1973 by Saunders College Publishing. All rights reserved.

Text Typeface: Times Roman
Compositor: Progressive Information Technologies
Publisher: John Vondeling
Developmental Editor: Margaret Crocker
Managing Editor: Carol Field
Project Editor: Linda Boyle
Copy Editor: Ellen Thomas
Manager of Art and Design: Carol Bleistine
Cover Designer: Katy Needle
Text Designer: Susan Blaker
Text Artwork: Vantage Art
Product Manager: Angus McDonald
Director of EDP: Tim Frelick
Production Manager: Joanne Cassetti

Printed in the United States of America

CHEMICAL PRINCIPLES IN THE LABORATORY, sixth edition

ISBN 0-03-005939-9

Library of Congress Catalog Card Number: 95-067413
  7890123 071 10 98765

This book was printed on acid-free recycled content paper, containing **MORE THAN 10% POSTCONSUMER WASTE**

Sixth Edition

CHEMICAL PRINCIPLES IN THE LABORATORY

Emil J. Slowinski
Professor of Chemistry
Macalester College
St. Paul, Minnesota

Wayne C. Wolsey
Professor of Chemistry
Macalester College
St. Paul, Minnesota

William L. Masterton
Professor of Chemistry
University of Connecticut
Storrs, Connecticut

Saunders Golden Sunburst Series
Saunders College Publishing
HARCOURT BRACE COLLEGE PUBLISHERS

Fort Worth   Philadelphia   San Diego   New York   Orlando   Austin
San Antonio   Toronto   Montreal   London   Sydney   Tokyo

Preface

In spite of its many successful theories, chemistry remains, and probably always will remain, an experimental science. Most of the research, both in universities and in industry, is done in the laboratory rather than in the office or computing room, and it behooves the young student of chemistry to devote a substantial portion of time to the experimental aspects of the subject. It is not easy to become a good experimentalist, and it must be admitted that some chemists are not always effective in the laboratory. Yet even those chemists who do not work full-time in the laboratory must be familiar with the available experimental methods, the proper design of experiments, and the interpretation of experimental results. As beginning chemists, students will find that their efforts in the laboratory will be rewarded by a better understanding of the concepts of chemistry as well as an appreciation of what is required in the way of technique and interpretation if they are to be able to find or demonstrate any chemically significant relations.

In writing this manual, the authors have attempted to illustrate many of the established principles of chemistry with experiments that are as interesting and challenging as possible. For the most part, the experimental procedures and methods for processing data are described in great detail, so that students of widely varying backgrounds and abilities will be able to see how to perform the experiments properly and how to interpret them. Many of the experiments in this manual have not appeared elsewhere, and were developed and tested in the general chemistry laboratories of Macalester College and the University of Connecticut. We have included more experiments than can be conveniently done in a year in the usual laboratory program, so instructors may select experiments in a flexible way to meet the needs of their particular courses. In many of the experiments, unknowns are assigned to students to ensure their working independently and to add a measure of realism to the laboratory experience.

In preparation for writing this sixth edition, we sent out questionnaires to current users of the manual, asking for comments about the experiments and suggestions for changes or improvements. These comments have been carefully considered, and many of the suggestions have been incorporated in this manual. Several of the experiments have improved procedures, a few new experiments have been added, and a few have been dropped. We have added optional microscale procedures for two experiments, and have added two appendices describing instruments and techniques for making measurements and some mathematical principles relevant to experimental results. Finally, in the last appendix we have included some suggestions for possible extension of the experiments to problems involving chemicals we encounter in our daily lives. The order of the experiments makes them compatible with the order of topics in the text *Chemistry: Principles and Reactions,* Second Edition, by Masterton and Hurley. We believe, however, that the overall set of experiments should be appropriate for use with most modern texts in general chemistry.

As in earlier editions, we include with each experiment an advance study assignment, designed to help the student prepare for the laboratory session. The questions in the advance study assignments include in nearly every case data similar to those the student will obtain in the experiment. Directions for treating the data have been given in great detail, so if the student properly completes the advance study assignments, he or she should have no trouble in performing the experiments or in processing the data.

We have selected the experiments with some regard to cost, since both chemicals and equipment are expensive. In the teacher's guide, we note the cost of the chemicals per student for each experiment. We have attempted to keep the experiments safe, have dropped toxic reagents or suspected carcinogens where that was feasible, and have included safety warnings in experimental procedures where they seemed indicated. At the end of the Teacher's Guide we have included directions for disposal of reaction products for each of the experiments.

The system of units we have used in this manual is essentially the same as in the fifth edition. For the most part, the system is SI, but we have kept mm Hg and atmospheres for pressure measurements, and usually use milliliters and liters for expressing volumes.

The authors gratefully acknowledge the assistance of the following persons: Antoinette Amegayibor, a Macalester student from Ghana, who tested the experimental procedures in the new and modified experiments and who made many helpful suggestions; Barbara Ekeberg, who took time from her summer vacation to do some typing on the manuscript and teacher's guide; and to Sherman Schultz, who generously agreed to take the photographs of Macalester chem lab equipment that appear in Appendix IV.

We would like to thank those who replied to our questionnaire for taking the time to give us their opinions and suggestions. We are grateful for the support of those who have used our earlier editions and would appreciate any comments you may have about this edition of the manual.

E. J. Slowinski

W. C. Wolsey

W. L. Masterton

Safety in the Laboratory

Read this section before performing any of the experiments in this manual.

A chemistry laboratory can be, and should be, a safe place in which to work. Yet each year in academic and industrial laboratories accidents occur that in some cases injure seriously, or kill, chemists. Most of these accidents could have been foreseen and prevented, had the chemists involved used the proper judgment and taken proper precautions.

The experiments you will be performing have been selected at least in part because they can be done safely. Instructions in the procedures should be followed carefully and in the order given. Sometimes even a change in concentration of one reagent is sufficient to change the conditions of a chemical reaction so as to make it occur in a different way, perhaps at a highly accelerated rate. So, do not deviate from the procedure given in the manual when performing experiments unless specifically told to do so by your instructor.

EYE PROTECTION. One of the simplest, and most important, things you can do in the laboratory to avoid injury is to protect your eyes by routinely wearing safety glasses. Your instructor will tell you what eye protection to use, and you should use it. Goggles worn up on the hair may be attractive, but they are not protective. It is not advisable to wear contact lenses in the laboratory.

CHEMICAL REAGENTS. Chemicals in general are toxic materials. This means that they can act as poisons or carcinogens (causes of cancer) if they get into your digestive or respiratory system. Never taste a chemical substance, and avoid getting any chemical on your skin. If that should happen, wash it off promptly with plenty of water. Also, wash your face and hands when you are through working in the laboratory. Do not pipet by mouth; when pipetting, use a rubber bulb or other device to suck up the liquid. Avoid breathing vapors given off by reagents or reactions. If directed to smell a vapor, do so cautiously. Use the hood when the directions call for it.

Some reagents, such as concentrated acids or bases, or bromine, are caustic, which means that they can cause chemical burns on your skin and eat through your clothing. Where such reagents are being used, we note the potential danger with a *CAUTION* sign at that point in the procedure. Be particularly careful when carrying out that step. Always read the label on a reagent bottle before using it; there is a lot of difference between the properties of 1 M H_2SO_4 and those of concentrated (18 M) H_2SO_4.

A few of the chemicals we use are flammable. These include hexane, ethanol, and acetone. Keep your Bunsen burner well away from any open beakers containing such chemicals, and be careful not to spill them on the laboratory bench, where they might easily get ignited.

When disposing of the chemical products from an experiment, use good judgment. Some dilute, nontoxic solutions can be poured down the sink and flushed with water. Insoluble or toxic materials should be put in the waste crocks provided for that purpose. Your lab instructor may give you instructions for treatment and disposal of the products from specific experiments.

SAFETY EQUIPMENT. In the laboratory there are various pieces of safety equipment, which may include a safety shower, an eye wash fountain, a fire extinguisher, and a fire blanket. Learn where these items are, so that you will not have to look all over if you ever need them in a hurry.

LABORATORY ATTIRE. Come to the laboratory in sensible clothing. Long, flowing robes are out, as are bare feet. Sandals and open-toed shoes offer less protection than regular shoes. Keep long hair tied back, out of the way of flames and reagents.

IF AN ACCIDENT OCCURS. During the laboratory course a few accidents will probably occur. For the most part these will not be serious, and might involve a spilled reagent, a beaker of hot water that gets tipped over, a dropped test tube, or a small fire.

A common response in such a situation is panic. A student may respond to an otherwise minor accident by doing something irrational, like running from the laboratory when the remedy for the accident is close at hand. If an accident happens to another student, watch for signs of panic and tell the student what to do; if it seems necessary, help him or her do it. Call the instructor for assistance.

Chemical spills are best handled by washing the area quickly with water from the nearest sink. Use the eye wash fountain if you get something in your eye. In case of a severe chemical spill on your clothing or shoes, use the emergency shower and take off the affected clothing. In case of a fire in a beaker, on a bench, or on your clothing or that of another student, do not panic and run. Smother the fire with an extinguisher, with a blanket, or with water, as seems most appropriate at the time. If the fire is in a piece of equipment or on the lab bench and does not appear to require instant action, have your instructor put the fire out. If you cut yourself on a piece of broken glass, tell your instructor, who will assist you in treating it.

A MESSAGE TO FOREIGN STUDENTS. Many students from foreign countries take courses in chemistry before they are completely fluent in English. If you are such a student, it may be that in some experiments you will be given directions that you do not completely understand. If that happens, do not try to do that part of the experiment by simply doing what the student next to you seems to be doing. Ask that student, or the instructor, what the confusing word or phrase means, and when you understand what you should do, go ahead. You will soon learn the language well enough, but until you feel comfortable with it, do not hesitate to ask others to help you with unfamiliar phrases and expressions.

Although we have spent considerable time here describing some of the things you should be concerned with in the laboratory from a safety point of view, this does not mean you should work in the laboratory in fear and trepidation. Chemistry is not a dangerous activity when practiced properly. Chemists as a group live longer than other professionals, in spite of their exposure to potentially dangerous chemicals. In this manual we have attempted to describe safe procedures and to employ chemicals that are safe when used properly. Many thousands of students have performed the experiments without having accidents, so you can too. However, we authors cannot be in the laboratory when you carry out the experiments to be sure that you observe the necessary precautions. You and your laboratory supervisor must, therefore, see to it that the experiments are done properly and assume responsibility for any accidents or injuries that may occur.

C ontents

Experiment 1
The Densities of Liquids and Solids 1

Experiment 2
Resolution of Matter into Pure Substances, I. Paper Chromatography 7

Experiment 3
Resolution of Matter into Pure Substances, II. Fractional Crystallization 15

Experiment 4
Determination of a Chemical Formula 23

Experiment 5
Identification of a Compound Using Mass Relationships 31

Experiment 6
Analysis of an Unknown Chloride 37

Experiment 7
Determination of the Barometric Pressure—Charles' Law 45

Experiment 8
Molar Mass of a Volatile Liquid 57

Experiment 9
Analysis of an Aluminum-Zinc Alloy 63

Experiment 10
The Atomic Spectrum of Hydrogen 73

Experiment 11
The Alkaline Earths and the Halogens—Two Families in the Periodic Table 83

Experiment 12
The Geometrical Structure of Molecules 91

Experiment 13
Heat Effects and Calorimetry 101

Experiment 14
The Vapor Pressure and Heat of Vaporization of Liquids 109

Experiment 15
The Structure of Crystals 119

Experiment 16
Classification of Chemical Substances 133

Experiment 17
Molar Mass Determination by Depression of the Freezing Point 141

Experiment 18
Rates of Chemical Reactions, I. The Iodination of Acetone 151

Experiment 19
Rates of Chemical Reactions, II. A Clock Reaction 161

Experiment 20
Properties of Systems in Equilibrium—Le Châtelier's Principle 173

Experiment 21
Determination of the Equilibrium Constant for a Reaction 185

Experiment 22
Standardization of a Basic Solution and the Determination of the Equivalent
Mass of an Acid 195

Experiment 23
pH—Buffers and Their Properties 203

Experiment 24
Determination of the Solubility Product of PbI_2 215

Experiment 25
Relative Stabilities of Complex Ions and Precipitates Prepared from Cu(II) 223

Experiment 26
Determination of the Hardness of Water 233

Experiment 27
Synthesis and Analysis of a Coordination Compound 241

Experiment 28
Determination of Iron by Reaction with Permanganate—A Redox Titration 253

Experiment 29
Determination of an Equivalent Mass by Electrolysis 259

Experiment 30
Voltaic Cell Measurements 267

Experiment 31
Preparation of Copper(I) Chloride 277

Experiment 32
Development of a Scheme for Qualitative Analysis 285

Experiment 33
Qualitative Analysis of Group I Cations 291

Experiment 34
Qualitative Analysis of Group II Cations 299

Experiment 35
Qualitative Analysis of Group III Cations 307

Experiment 36
The Ten Test Tube Mystery 315

Experiment 37
Laboratory Examination on Qualitative Analysis of Cations 323

Experiment 38
Some Nonmetals and Their Compounds—Preparation and Properties 327

Experiment 39
Spot Tests for Some Common Anions 335

Experiment 40
Sulfur Chemistry 343

Experiment 41
Preparation of Aspirin 351

Experiment 42
Analysis for Vitamin C 359

Appendix I
Vapor Pressure of Water 365

Appendix II
Summary of Solubility Properties of Ions and Solids 367

Appendix III
Table of Atomic Masses 369

Appendix IV
Making Measurements—Laboratory Techniques 371

Appendix V
Mathematical Considerations—Making Graphs 383

Appendix VI
Suggested Locker Equipment 389

Appendix VII
Suggestions for Extension of the Experiments to "Real World Problems" 391

Experiment 1

THE DENSITIES OF LIQUIDS AND SOLIDS

One of the fundamental properties of any sample of matter is its density, which is its mass per unit of volume. The density of water is exactly 1.00000 g/cm³ at 4°C and is slightly less than one at room temperature (0.9970 g/cm³ at 25°C). Densities of liquids and solids range from values less than that of water to values considerably greater than that of water. Osmium metal has a density of 22.5 g/cm³ and is probably the densest material known at ordinary pressures.

In any density determination, two quantities must be determined—the mass and the volume of a given quantity of matter. The mass can easily be determined by weighing a sample of the substance on a balance. The quantity we usually think of as "weight" is really the mass of a substance. In the process of "weighing" we find the mass, taken from a standard set of masses, that experiences the same gravitational force as that experienced by the given quantity of matter we are weighing. The mass of a sample of liquid in a container can be found by taking the difference between the mass of the container plus the liquid and the mass of the empty container.

The volume of a liquid can easily be determined by means of a calibrated container. In the laboratory a graduated cylinder is often used for routine measurements of volume. Accurate measurement of liquid volume is made by using a pycnometer, which is simply a container having a precisely definable volume. The volume of a solid can be determined by direct measurement if the solid has a regular geometrical shape. Such is not usually the case, however, with ordinary solid samples. A convenient way to determine the volume of a solid is to measure accurately the volume of liquid displaced when an amount of the solid is immersed in the liquid. The volume of the solid will equal the volume of liquid which it displaces.

In this experiment we will determine the density of a liquid and a solid by the procedure we have outlined. First we weigh an empty flask and its stopper. We then fill the flask completely with water, measuring the mass of the filled stoppered flask. From the difference in these two masses we find the mass of water and then, from the known density of water, we determine the volume of the flask. We empty and dry the flask, fill it with an unknown liquid, and weigh again. From the mass of the liquid and the volume of the flask we find the density of the liquid. To determine the density of an unknown solid metal, we add the metal to the dry empty flask and weigh. This allows us to find the mass of the metal. We then fill the flask with water, leaving the metal in the flask, and weigh again. The increase in mass is that of the added water; from that increase, and the density of water, we calculate the volume of water we added. The volume of the metal must equal the volume of the flask minus the volume of water. From the mass and volume of the metal we calculate its density. The calculations involved are outlined in detail in the Advance Study Assignment.

Experimental Procedure

A. Mass of a Coin

After you have been shown how to operate the analytical balances in your laboratory, read the section on balances in Appendix IV. Take a coin and measure its mass to 0.0001 g. Record the mass on the DATA page. If your balance has a TARE bar, use it to re-zero the balance. Take another coin and weigh it, recording its mass. Remove both coins, zero the balance, and weigh both coins together, recording the total mass. If you have no TARE bar on your balance, add the second coin and measure and record the mass of the two coins. Then remove both coins and find the mass of the second one by itself. When you are satisfied that your results are those you would expect, go to the stockroom and obtain a glass-stoppered flask, which will serve as a pycnometer, and samples of an unknown liquid and an unknown metal.

B. Density of a Liquid

If your flask is not clean and dry, clean it with soap and water, rinse it with a few cubic centimeters of acetone, and dry it by letting it stand for a few minutes in the air or by *gently* blowing compressed air into it for a few moments.

Weigh the dry flask with its stopper on the analytical balance, or the toploading balance if so directed, to the nearest milligram. Fill the flask with distilled water until the liquid level is nearly to the *top* of the ground surface in the neck. Put the stopper in the flask in order to drive out *all* the air and any excess water. Work the stopper gently into the flask, so that it is firmly seated in position. Wipe any water form the outside of the flask with a towel and soak up all excess water from around the top of the stopper.

Again weigh the flask, which should be completely dry on the outside and full of water, to the nearest milligram. Given the density of water at the temperature of the laboratory and the mass of water in the flask, you should be able to determine the volume of the flask very precisely. Empty the flask, dry it, and fill it with your unknown liquid. Stopper and dry the flask as you did when working with the water and then weigh the stoppered flask full of the unknown liquid, making sure its surface is dry. This measurement, used in conjunction with those you made previously, will allow you to find accurately the density of your unknown liquid.

C. Density of a Solid

Pour your sample of liquid from the flask into its container. Rinse the flask with a small amount of acetone and dry it thoroughly. Add small chunks of the metal sample to the flask until the flask is at least half full. Weigh the flask, with its stopper and the metal, to the nearest milligram. You should have at least 50 g of metal in the flask.

Leaving the metal in the flask, fill the flask with water and then replace the stopper. Roll the metal around in the flask to make sure that no air remains between the metal pieces. Refill the flask if necessary, and then weigh the dry, stoppered flask full of water plus the metal sample. Properly done, the measurements you have made in this experiment will allow a calculation of the density of your metal sample that will be accurate to about 0.1%.

Pour the water from the flask. Put the metal in its container. Dry the flask and return it with its stopper and your metal sample, along with the sample of unknown liquid, to the stockroom.

Data and Calculations: Densities of Liquids and Solids

A. Mass of coin 1 _____ g Mass of coin 2 _____ g

Mass of coins 1 and 2 weighed together _____ g
What general law is illustrated by the results of this experiment?

B. Density of unknown liquid

Mass of empty flask plus stopper _____ g

Mass of stoppered flask plus water _____ g

Mass of stoppered flask plus liquid _____ g

Mass of water _____ g

Volume of flask (density of H_2O at 25°C, 0.9970 g/cm³; at
20°C, 0.9982 g/cm³) _____ cm³

Mass of liquid _____ g

Density of liquid _____ g/cm³

To how many significant figures can the liquid density be properly reported? (See Appendix V.) _____

C. Density of unknown metal

Mass of stoppered flask plus metal _____ g

Mass of stoppered flask plus metal plus water _____ g

Mass of metal _____ g

Mass of water _____ g

Volume of water _____ cm³

Volume of metal _____ cm³

(continued on following page)

(continued)

Density of metal   _____ g/cm^3

To how many significant figures can the density of the metal be properly reported?   _____

Explain why the value obtained for the density of the metal is likely to have a larger percentage error than that found for the liquid.

Unknown liquid no. _____   Unknown solid no. _____

Advance Study Assignment: Densities of Solids and Liquids

The advance study assignments in this laboratory manual are designed to assist you in making the calculations required in the experiment you will be doing. We do this by furnishing you with sample data and showing in some detail how that data can be used to obtain the desired results. In the advance study assignments we will often include the guiding principles as well as the specific relationships to be employed. If you work through the steps in each calculation by yourself, you should have no difficulty when you are called upon to make the necessary calculations on the basis of the data you obtain in the laboratory.

1. *Finding the volume of a flask.* A student obtained a clean, dry glass-stoppered flask. She weighed the flask and stopper on an analytical balance and found the total mass to be 32.634 g. She then filled the flask with water and obtained a mass for the full stoppered flask of 59.479 g. From these data, and the fact that at the temperature of the laboratory the density of water was 0.9973 g/cm³, find the volume of the stoppered flask.

   a. First we need to obtain the mass of the water in the flask. This is found by recognizing that the mass of a sample is equal to the sum of the masses of its parts. For the filled stoppered flask:

      Mass of filled stoppered flask = mass of empty stoppered flask + mass of water, so mass of water = mass of filled flask − mass of empty flask

      Mass of water = _____ g − _____ g = _____ g

      Many mass and volume measurements in chemistry are made by the method used in 1a. This method is called measuring by difference, and is a very useful one.

   b. The density of a pure substance is equal to its mass divided by its volume:

      $$\text{Density} = \frac{\text{mass}}{\text{volume}} \quad \text{or} \quad \text{volume} = \frac{\text{mass}}{\text{density}}$$

      The volume of the flask is equal to the volume of the water it contains. Since we know the mass and density of the water, we can find its volume and that of the flask. Make the necessary calculation.

      Volume of water = volume of flask = _____ cm³

2. *Finding the density of an unknown liquid.* Having obtained the volume of the flask, the student emptied the flask, dried it, and filled it with an unknown whose density she wished to determine. The mass of the stoppered flask when completely filled with liquid was 50.376 g. Find the density of the liquid.

   a. First we need to find the mass of the liquid by measuring by difference:

      Mass of liquid = _____ g − _____ g = _____ g

      (continued on following page)

b. Since the volume of the liquid equals that of the flask, we know both the mass and volume of the liquid and can easily find its density using the equation in 1b. Make the calculation.

Density of liquid = _____ g/cm³

3. *Finding the density of a solid.* The student then emptied the flask and dried it once again. To the empty flask she added pieces of a metal until the flask was about half full. She weighed the stoppered flask and its metal contents and found that the mass was 152.047 g. She then filled the flask with water, stoppered it, and obtained a total mass of 165.541 g for the flask, stopper, metal, and water. Find the density of the metal.

a. To find the density of the metal we need to know its mass and volume. We can easily obtain its mass by the method of differences:

Mass of metal = _____ g − _____ g = _____ g

b. To determine the volume of metal, we note that the volume of the flask must equal the volume of the metal plus the volume of water in the filled flask containing both metal and water. If we can find the volume of water, we can obtain the volume of metal by the method of differences. To obtain the volume of the water we first calculate its mass:

Mass of water = mass of (flask + stopper + metal + water)
$$- \text{ mass of (flask + stopper + metal)}$$

Mass of water = _____ g − _____ g = _____ g

The volume of water is found from its density, as in 1b. Make the calculation.

Volume of water = _____ cm³

c. From the volume of the water we calculate the volume of metal:

Volume of metal = volume of flask − volume of water

Volume of metal = _____ cm³ − _____ cm³ = _____ cm³

From the mass of and volume of metal we find the density, using the equation in 1b. Make the calculation.

Density of metal = _____ g/cm³

Now go back to Question 1 and check to see that you have reported the proper number of significant figures in each of the results you calculated in this assignment. Use the rules on significant figures as given in your chemistry text or Appendix V.

RESOLUTION OF MATTER INTO PURE SUBSTANCES, I. PAPER CHROMATOGRAPHY

The fact that different substances have different solubilities in a given solvent can be used in several ways to effect a separation of substances from mixtures in which they are present. We will see in an upcoming experiment how fractional crystallization allows us to obtain pure substances by relatively simple procedures based on solubility properties. Another widely used resolution technique, which also depends on solubility differences, is chromatography.

In the chromatographic experiment a mixture is deposited on some solid adsorbing substance, which might consist of a strip of filter paper, a thin layer of silica gel on a piece of glass, some finely divided charcoal packed loosely in a glass tube, or even some microscopic glass beads coated thinly with a suitable adsorbing substance and contained in a piece of copper tubing.

The components of a mixture are adsorbed on the solid to varying degrees, depending on the nature of the component, the nature of the adsorbent, and the temperature. A solvent is then caused to flow through the adsorbent solid under applied or gravitational pressure or by the capillary effect. As the solvent passes the deposited sample, the various components tend, to varying extents, to be dissolved and swept along the solid. The rate at which a component will move along the solid depends on its relative tendency to be dissolved in the solvent and adsorbed on the solid. The net effect is that, as the solvent passes slowly through the solid, the components separate from each other and move along as rather diffuse zones. With the proper choice of solvent and adsorbent, it is possible to resolve many complex mixtures by this procedure. If necessary, we can usually recover a given component by identifying the position of the zone containing the component, removing that part of the solid from the system, and eluting the desired component with a suitable good solvent.

The name given to a particular kind of chromatography depends upon the manner in which the experiment is conducted. Thus, we have column, thin-layer, paper, and vapor chromatography, all in very common use. Chromatography in its many possible variations offers the chemist one of the best methods, if not the best method, for resolving a mixture into pure substances, regardless of whether that mixture consists of a gas, a volatile liquid, or a group of nonvolatile, relatively unstable, complex organic compounds.

In this experiment we will use paper chromatography to separate a mixture of metallic ions in solution. A sample containing a few micrograms of ions is applied as a spot near one edge of a piece of filter paper. That edge is immersed in a solvent, with the paper held vertically. As the solvent rises up the paper by capillary action, it will carry the metallic ions along with it to a degree that depends upon the relative tendency of each ion to dissolve in the solvent and adsorb on the paper. Because the ions differ in their properties, they move at different rates and become separated on the paper. The position of each ion during the experiment can be recognized if the ion is colored, as some of them are. At the end of the experiment their positions are established more clearly by treating the paper with a staining reagent which reacts with each ion to produce a colored product. By observing the position and color of the spot produced by each ion, and the positions of the spots produced by an unknown containing some of those ions, you can readily determine the ions present in the unknown.

It is possible to describe the position of spots such as those you will be observing in terms of a quantity called the R_f value. In the experiment the solvent rises a certain distance, say L centimeters. At the same time a given component will usually rise a smaller distance, say D centimeters. The ratio of D/L is called the R_f value for that component:

$$R_f = \frac{D}{L} = \frac{\text{distance component moves}}{\text{distance solvent moves}} \qquad (1)$$

The R_f value is a characteristic property of a given component in a chromatography experiment conducted under particular conditions. It does not depend upon concentration or upon the other components present. Hence it can be reported in the literature and used by other researchers doing similar analyses. In the experiment you will be doing, you will be asked to calculate the R_f values for each of the cations studied.

Experimental Procedure

From the stockroom obtain an unknown and a piece of filter paper about 19 cm long and 11 cm wide. Along the 19-cm edge, draw a pencil line about 1 cm from that edge. Starting 1.5 cm from the end of the line, mark the line at 2-cm intervals. Label the segments of the line as shown in Figure 2.1, with the formulas of the ions to be studied and the known and unknown mixtures.

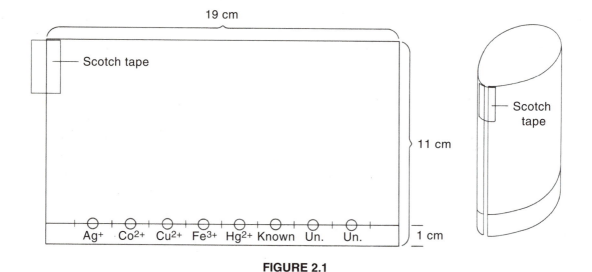

FIGURE 2.1

Put two or three drops of 0.1 M solutions of the following compounds in small micro test tubes, one solution to a tube:

$$AgNO_3 \quad Co(NO_3)_2 \quad Cu(NO_3)_2 \quad Fe(NO_3)_3 \quad Hg(NO_3)_2$$

In solution these substances exist as ions. The metallic cations are Ag^+, Co^{2+}, Cu^{2+}, Fe^{3+}, and Hg^{2+} respectively. One drop of each solution contains about 50 micrograms of cation. Into a sixth micro test tube put two drops of each of the five solutions; swirl until the solutions are well mixed. This mixture will be our known, since we know it contains all of the cations.

Your instructor will furnish you with a fine capillary tube, which will serve as an applicator. Test the application procedure by dipping the applicator into one of the colored solutions and touching it momentarily to a round piece of filter paper. The liquid from the applicator should form a spot no larger than 8 mm in diameter. Practice making spots until you can reproduce the spot size each time.

Clean the applicator by dipping it about 1 cm into distilled water and then touching the round filter paper to remove the liquid. Continue contact until all the liquid in the tube is gone. Repeat the cleaning procedure one more time. Dip the applicator into one of the cation solutions and put a spot on the line on the rectangular filter paper in the region labeled for that cation. Clean the applicator twice, and repeat the procedure with another solution. Continue this approach until you have put a spot for each of the five cations and the known and unknown on the paper, cleaning the applicator between solutions. Dry the paper by moving it in the air or holding it briefly in front of a hair dryer or heat lamp (low setting). Apply the known and unknown three more times to the same spots; the known and unknown are less concentrated than the cation solutions, so this procedure will increase the amount of each ion in the spots. Make sure that you dry the spots between applications, since otherwise they will get larger. Don't heat the paper more than necessary, just enough to dry the spots.

Draw about 15 mL of eluting solution from the supply on the reagent shelf. This solution is made by mixing a solution of HCl, hydrochloric acid, with ethanol and butanol, which are organic solvents. Pour the eluting solution into a 600-mL beaker and cover with a watch glass.

Check to make sure that the spots on the filter paper are all dry. Place a 4- to 5-cm length of Scotch tape along the upper end of the left edge of the paper, as shown in Figure 2.1, so that about half of the tape is on the paper. Form the paper into a cylinder by attaching the tape to the other edge, in such a way that the edges are parallel but do not overlap. When you are finished, the pencil line at the bottom of the cylinder should form a circle, approximately anyway, and the two edges of the paper should not quite touch. Stand the cylinder up on the lab bench to check that such is the case and readjust the tape if necessary. *Do not* tape the lower edges of the paper together.

Place the cylinder in the eluting solution in the 600-mL beaker, with the sample spots down near the liquid surface. The paper should not touch the wall of the beaker. Cover the beaker with the watch glass. The solvent will gradually rise by capillary action up the filter paper, carrying along the cations at different rates. After the process has gone on for a few minutes, you should be able to see colored spots on the paper, showing the positions of some of the cations.

While the experiment is proceeding, you can test the effect of the staining reagent on the different cations. Put an 8-mm spot of each of the cation solutions on a clean piece of round filter paper, labeling each spot and cleaning the applicator between solutions. Dry the spots as before. Some of them will have a little color; record those colors on the Data sheet. Put the filter paper on a paper towel, and, using the spray bottle on the lab bench, spray the paper evenly with the staining reagent, getting the paper moist but not really wet. The staining reagent is a solution containing potassium ferrocyanide and potassium iodide. This reagent forms colored precipitates or reaction products with many cations, including all of those used in this experiment. Note the colors obtained with each of the cations. Considering that each spot contains less than 50 micrograms of cation, the tests are quite definitive.

When the eluting solution has risen to within about 2 cm of the top of the filter paper (it will take about 75 minutes), remove the cylinder from the beaker and take off the tape. Draw a pencil line along the solvent front. Dry the paper with gentle heat until it is quite dry. Note any cations that must be in your unknown by virtue of your being able to see their colors. Then, with the paper on a paper towel, spray it as before with the staining reagent. Any cations you identified in your unknown before staining should be observed, as well as any that require staining for detection.

Measure the distance from the straight line on which you applied the spots to the solvent front, which is distance L in Equation 1. Then measure the distance from the pencil line to the center of the spot made by each of the cations, when pure and in the known; this is distance D. Calculate R_f value for each cation. Then calculate R_f values for the cations in the unknown. How do the R_f values compare?

When you are finished with the experiment, pour the eluting solution into the waste crock, not down the sink. Wash your hands before leaving the laboratory.

Data and Calculations: Resolution of Matter into Pure Substances, I. Paper Chromatography

Colors (if observed)	Ag^+	Co^{2+}	Cu^{2+}	Fe^{3+}	Hg^{2+}
Dry	_____	_____	_____	_____	_____
After staining	_____	_____	_____	_____	_____
Distance solvent moved *(L)*	_____	_____	_____	_____	_____
Distance cation moved *(D)*	_____	_____	_____	_____	_____
R_f	_____	_____	_____	_____	_____

Known Mixture

	Ag^+	Co^{2+}	Cu^{2+}	Fe^{3+}	Hg^{2+}
Distance solvent moved	_____	_____	_____	_____	_____
Distance cation moved	_____	_____	_____	_____	_____
R_f	_____	_____	_____	_____	_____

Unknown Mixture

Cations identified

	Ag^+	Co^{2+}	Cu^{2+}	Fe^{3+}	Hg^{2+}
Dry	_____	_____	_____	_____	_____
After staining	_____	_____	_____	_____	_____
Distance solvent moved	_____	_____	_____	_____	_____
Distance cation moved	_____	_____	_____	_____	_____
R_f	_____	_____	_____	_____	_____
Composition of unknown	_____	_____	_____	_____	_____

Unknown no. _____

Advance Study Assignment: Resolution of Matter into Pure Substances, I. Paper Chromatography

1. A student chromatographs a mixture, and after developing the spots with a suitable reagent he observes the following:

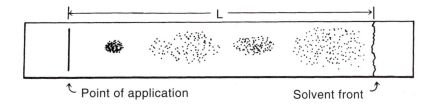

What are the R_f values for the four spots?

2. Explain, in your own words, why samples can often be separated into their components by chromatography.

3. The solvent moves 3 cm in about 5 minutes. Why shouldn't the experiment be stopped at that time instead of waiting 75 minutes for the solvent to move 10 cm?

4. In this experiment it takes about 10 microliters of solution to produce a spot 1 cm in diameter. If the $Cu(NO_3)_2$ solution contains about 6 g Cu^{2+} per liter, how many micrograms of Cu^{2+} ion are there in one spot?

_____ micrograms

Experiment 3

RESOLUTION OF MATTER INTO PURE SUBSTANCES, II. FRACTIONAL CRYSTALLIZATION

One of the important problems faced by chemists is that of determining the nature and state of purity of the substances with which they work. In order to perform meaningful experiments, chemists must ordinarily use essentially pure substances, which are often prepared by separation from complex mixtures.

In principle the separation of a mixture into its component substances can be accomplished by carrying the mixture through one or more physical changes, experimental operations in which the nature of the components remains unchanged. Because the physical properties of various pure substances are different, physical changes frequently allow an enrichment of one or more substances in one of the fractions that is obtained during the change. Many physical changes can be used to accomplish the resolution of a mixture, but in this experiment we will restrict our attention to one of the simpler ones in common use, namely, fractional crystallization.

The solubilities of solid substances in different kinds of liquid solvents vary widely. Some substances are essentially insoluble in all known solvents; the materials we classify as macromolecular are typical examples. Most materials are noticeably soluble in one or more solvents. Those substances that we call salts often have very appreciable solubility in water but relatively little solubility in any other liquids. Organic compounds, whose molecules contain carbon and hydrogen atoms as their main constituents, are often soluble in organic liquids such as benzene or carbon tetrachloride.

We also often find that the solubility of a given substance in a liquid is sharply dependent on temperature. Most substances are more soluble in a given solvent at high temperatures than at low temperatures, although there are some materials whose solubility is practically temperature-independent and a few others that become less soluble as temperature increases.

By taking advantage of the differences in solubility of different substances we often find it possible to separate the components of a mixture in essentially pure form.

In this experiment you will be given a sample containing silicon carbide, potassium nitrate, and copper sulfate. Your problem will be to separate two pure components from the mixture, using water as the solvent. Silicon carbide, SiC, is a black, very hard material; it is the classic abrasive, and completely insoluble in water. Potassium nitrate, KNO_3, and copper sulfate, $CuSO_4 \cdot 5 H_2O$, are water-soluble ionic substances, with different solubilities at different temperatures, as indicated in Figure 3.1. The copper sulfate we will use is blue in its crystalline hydrate and in solution. The solubility of the hydrate increases fairly rapidly with temperature. Potassium nitrate is a white solid, colorless in solution. Its solubility increases about 20-fold between 0°C and 100°C.

Given a mixture containing roughly equal amounts of SiC and KNO_3 and a small amount of $CuSO_4 \cdot 5 H_2O$, we separate out the silicon carbide first. This is done by simply stirring the mixture with water, which dissolves all of the potassium nitrate and copper sulfate in the mixture. The insoluble silicon carbide remains behind and is filtered off.

The solution obtained after filtration contains KNO_3 and $CuSO_4$ in a rather large amount of water. Some of the water is removed by boiling, and then the solution is cooled to 0°C. At

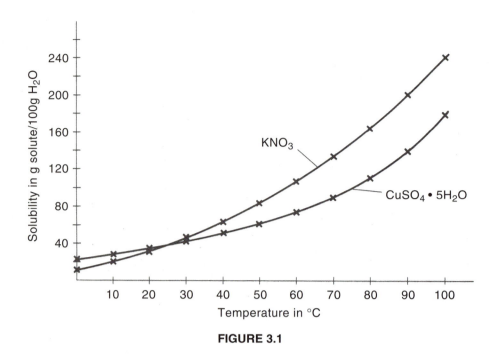

FIGURE 3.1

that point the KNO_3 is not very soluble, and most of it crystallizes from solution. Since $CuSO_4$ is not present in large amount, its solubility is not exceeded and it remains in solution. The solid KNO_3 is separated from the solution by filtration. This procedure, by which a substance can be separated from an impurity, is called fractional crystallization.

The solid potassium nitrate one recovers is contaminated by a small amount of copper sulfate. The purity of the solid can be markedly increased by stirring it with a small amount of water and then filtering off the dissolved $CuSO_4$. The purity can be established by the intensity of the color produced by the copper impurity when treated with ammonia, NH_3.

Experimental Procedure

WEAR YOUR SAFETY GLASSES WHILE
PERFORMING THIS EXPERIMENT

Obtain from the stockroom a Buchner funnel, a suction flask, and a sample (about 20 grams) of your unknown solid mixture.

Weigh a 150-mL beaker (to ±0.1 g) on a top-loading balance. Add the sample and weigh again. Then add about 40 mL of distilled water, which will be enough to dissolve the soluble solids. Light your Bunsen burner and adjust the flame so it is blue, quiet, and of moderate size.

Separation of SiC

Support the beaker with its contents on a piece of wire gauze on an iron ring. Warm gently to about 50°C, while stirring the mixture. When the blue and white solids are all dissolved, pour the contents of the beaker into a Buchner funnel while gentle suction is being applied (see Fig. 3.2 and Appendix IV). Transfer as much of the black solid carbide as you can to the funnel, using your rubber policeman. Transfer the blue filtrate to the (cleaned) 150-mL beaker and add 15 drops of 6 M HNO_3, nitric acid, which will help ensure that the copper sulfate remains in solution in later steps. Re-assemble the funnel, apply suction, and wash the SiC on the filter paper with distilled water. Continue the suction for a few minutes to dry the SiC. Turn off the suction, and, using your spatula, lift the filter paper and the SiC crystals from the funnel, and put the paper on the lab bench so that the crystals may dry in the air. Prepare some ice-cold distilled water by putting your wash bottle in an ice-water bath.

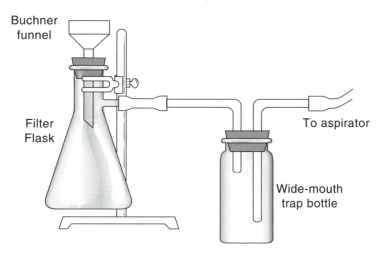

FIGURE 3.2 To operate the Buchner funnel, put a piece of circular filter paper in the funnel. Turn on suction and spray filter paper with distilled water from wash bottle. Keep suction on while filtering sample.

Separation of KNO_3

Heat the blue filtrate in the beaker to the boiling point, and then boil gently until the white crystals of KNO_3 are visible in the liquid.

C A U T I O N: *The hot liquid may have a tendency to bump, so do not heat it too strongly.*

When white crystals are clearly apparent (the solution may appear cloudy at that point) stop heating and add 12 mL distilled water to the solution. Stir the mixture with a glass stirring rod to dissolve the solids, including any on the wall; if necessary, warm the solution but do not boil it.

Cool the solution to room temperature in a water bath, and then to about 0°C in an ice bath. White crystals of KNO_3 will come out of solution. Stir the cold slurry of crystals for several minutes. Check the temperature of the slurry with your thermometer. It should be no more than 3°C. Continue stirring until your mixture gets to that temperature or even a bit lower. Assemble the Buchner funnel; chill it by adding 100 mL ice-cold distilled water and, after about a minute, drawing the water through with suction. Filter the KNO_3 slurry through the cold Buchner funnel. Your rubber policeman will be helpful when transferring the last of the crystals. Press the crystals dry with a clean piece of filter paper, and continue to apply suction for about 30 seconds. Turn off the suction. Lift the filter paper and the crystals from the funnel, and put the crystals on a dry piece of filter paper on the lab bench.

By this procedure you have separated most of the KNO_3 in your sample from the $CuSO_4$, which is present in the solution in the suction flask. This solution may now be discarded, so dispose of it as directed by your instructor.

Analysis of the Purity of the KNO_3

The KNO_3 crystals you have prepared contain a small amount of $CuSO_4$ as an impurity. To find the amount of $CuSO_4$ present, weigh out 0.5 g ($\pm$0.1 g) of the crystals into a weighed 50-mL beaker. Dissolve the crystals in 3 mL distilled water, and then add 3 mL 6 M NH_3, ammonia. The copper impurity will form a blue solution in the NH_3. Pour the solution into a small test tube. Compare the intensity of the blue color with that in a series of standard solutions prepared by your instructor. Estimate the relative concentration of $CuSO_4 \cdot 5 H_2O$ in your product.

Recrystallization of the KNO$_3$

Clean and dry the 150-mL beaker you used in the first part of the experiment, and to it add your sample of KNO$_3$. Weigh to ±0.1 g on the top-loading balance. Given the mass of KNO$_3$, use Figure 3.1 to estimate the amount of water needed to dissolve the solid at 100°C. Add three times that amount of distilled water to the sample. Stir the crystals for a minute or two to ensure that all the CuSO$_4$ impurity goes into solution.

Cool the mixture in an ice bath. This will recrystallize most of the KNO$_3$ that dissolved. After stirring for several minutes, check the temperature of the mixture with your thermometer. Continue cooling until the temperature is below 3°C. Then filter the slurry through an ice-cold Buchner funnel, transferring as much of the solid as possible to the funnel. Press the crystals dry with a piece of filter paper and continue to apply suction for a minute or so. Lift the filter paper from the funnel and put it with its batch of crystals on the lab bench. Weigh a piece of dry filter paper, to ±0.1 g, and then transfer the purified crystals to that paper. Weigh the paper and the crystals to ±0.1 g.

Determine the amount of CuSO$_4$ impurity in your recrystallized sample as you did with the first batch, and record that value. The recrystallization should have very significantly increased the purity of your KNO$_3$.

Weigh your dry SiC on its piece of filter paper, to ±0.1 g. Show your samples of SiC and KNO$_3$ to your instructor for evaluation. Then turn the samples in to the stockroom.

Data and Calculations: Resolution of Pure Substances, II. Fractional Crystallization

	Original Sample	*Recrystallized Sample*
Unknown no. _____		
Mass of 150-mL beaker	_____ g	
Mass of sample plus beaker	_____ g	
Mass of sample	_____ g	
Mass of 50-mL beaker	_____ g	
Mass of 50-mL beaker plus KNO_3	_____ g	
Mass of KNO_3 used in analysis	_____ g	
Percentage of $CuSO_4 \cdot 5 H_2O$ present in KNO_3	_____ %	
Mass of 150-mL beaker plus KNO_3	_____ g	
Mass of KNO_3 recovered (add 0.5 g)	_____ g	
Mass of filter paper		_____ g
Mass of paper plus purified KNO_3		_____ g
Mass of purified KNO_3 recovered		_____ g
Mass of 50-mL beaker plus purified KNO_3		_____ g
Mass of KNO_3 used in analysis		_____ g
Percentage of $CuSO_4 \cdot 5 H_2O$ in purified KNO_3		_____ %
Mass of SiC plus filter paper	_____ g	
Mass of SiC in sample	_____ g	
Percentage of SiC in sample	_____ %	
Percentage of sample recovered as KNO_3	_____ %	
Percentage of sample recovered as pure KNO_3		_____ %

Advance Study Assignment: Resolution of Matter into Pure Substances, II. Fractional Crystallization

1. Using Figure 3.1, determine

    a. the number of grams of KNO_3 that will dissolve in 100 g of H_2O at 100°C. If you need to, see Appendix V for a discussion of how to interpret a graph.

    _____ g KNO_3

    b. the number of grams of water required to dissolve 20 g of KNO_3 at 100°C. (Hint: your answer to 1a gives you the needed conversion factor for g KNO_3 to g H_2O.)

    _____ g H_2O

    c. the number of grams of water required to dissolve 2.0 g $CuSO_4 \cdot 5 H_2O$ at 100°C.

    _____ g H_2O

    d. the number of grams of water required at 100°C to dissolve a mixture containing 20 g KNO_3 and 2.0 g $CuSO_4 \cdot 5 H_2O$, assuming that the solubility of one substance is not affected by the presence of another.

    _____ g H_2O

2. To the solution in Problem 1d at 100°C, 15 g of water are added, and the solution is cooled to 0°C.

    a. How much KNO_3 remains in solution? (See Fig. 3.1.)

    _____ g KNO_3

    b. How much KNO_3 crystallizes out?

    _____ g KNO_3

    c. How much $CuSO_4 \cdot 5 H_2O$ crystallizes out?

    _____ g $CuSO_4 \cdot 5 H_2O$

    d. What percent of the KNO_3 in the sample is recovered?

    _____ %

Experiment 4

DETERMINATION OF A CHEMICAL FORMULA

When atoms of one element combine with those of another, the combining ratio is typically an integer or a simple fraction; $1:2$, $1:1$, $2:1$, and $2:3$ are ratios one might encounter. The simplest formula of a compound expresses that atom ratio. Some substances with the ratios we listed include $CaCl_2$, KBr, Ag_2O, and Fe_2O_3. When more than two elements are present in a compound, the formula still indicates the atom ratio. Thus the substance with the formula Na_2SO_4 indicates that the sodium, sulfur, and oxygen atoms occur in that compound in the ratio $2:1:4$. Many compounds have more complex formulas than those we have noted, but the same principles apply.

To find the formula of a compound we need to find the mass of each of the elements in a weighed sample of that compound. For example, if we resolved a sample of the compound NaOH weighing 40 grams into its elements, we would find that we obtained just about 23 grams of sodium, 16 grams of oxygen, and 1 gram of hydrogen. Since the atomic mass scale tells us that sodium atoms have a relative mass of 23, oxygen atoms a relative mass of 16, and hydrogen atoms a relative mass of just about 1, we would conclude that the sample of NaOH contained equal numbers of Na, O, and H atoms. Since that is the case, the atom ratio Na:O:H is $1:1:1$, and so the simplest formula is NaOH. In terms of moles, we can say that that one mole of NaOH, 40 grams, contains one mole of Na, 23 grams, one mole of O, 16 grams, and one mole of H, 1 gram, where we define the mole to be that mass in grams equal numerically to the sum of the atomic masses in an element or a compound. From this kind of argument we can conclude that the atom ratio in a compound is equal to the mole ratio. We get the mole ratio from chemical analysis, and from that the formula of the compound.

In this experiment we will use these principles to find the formula of the compound with the general formula $Cu_xCl_y \cdot zH_2O$, where the x, y, and z are integers which, when known, establish the formula of the compound. (In expressing the formula of a compound like this one, where water molecules remain intact within the compound, we retain the formula of H_2O in the formula of the compound.)

The compound we will study, which is called copper chloride hydrate, turns out to be ideal for one's first venture into formula determination. It is stable, can be obtained in pure form, has a characteristic blue-green color which changes as the compound is changed chemically, and is relatively easy to decompose into the elements and water. In the experiment we will first drive out the water, which is called the water of hydration, from an accurately weighed sample of the compound. This occurs if we gently heat the sample to a little over 100°C. As the water is driven out, the color of the sample changes from blue-green to a tan-brown color similar to that of tobacco. The compound formed is anhydrous (no water) copper chloride. If we subtract its mass from that of the hydrate, we can determine the mass of the water that was driven off, and, using the molar mass of water, find the number of moles of H_2O that were in the sample.

In the next step we need to find either the mass of copper or the mass of chlorine in the anhydrous sample we have prepared. It turns out to be much easier to determine the mass of the copper, and find the mass of chlorine by difference. We do this by dissolving the anhydrous sample in water, which gives us a green solution containing copper and chloride ions. To that solution we add some aluminum metal wire. Aluminum is what we call an active metal; in contact with a solution containing copper ions, the aluminum metal will react chemically with those ions, converting them to copper metal. The aluminum is said to reduce the copper ions to the metal, and is itself oxidized. The copper metal appears on the wire as the reaction proceeds, and has the typical red-orange color. When the reaction is complete, we remove the excess Al, separate the copper from the solution, and weigh the dried metal. From its mass we can calculate the number of moles of copper in the sample. We find the mass of chlorine by subtracting the mass of copper from that of the anhydrous copper chloride, and from that value determine the number of moles of chlorine. The mole ratio for $Cu:Cl:H_2O$ gives us the formula of the compound.

Experimental Procedure

Weigh a clean, dry crucible, without a cover, accurately on the analytical balance. Place about 1 gram of the unknown hydrated copper chloride in the crucible. With your spatula, break up any sizeable crystal particles by pressing them against the wall of the crucible. Then weigh the crucible and its contents accurately. Enter your results on the DATA page.

Place the uncovered crucible on a clay triangle supported by an iron ring. Light your bunsen burner away from the crucible, and adjust the burner so that you have a small flame. Holding the burner in your hand, gently heat the crucible as you move the burner back and forth. Do not overheat the sample. As the sample warms, you will see that the green crystals begin to change to brown around the edges. Continue gentle heating, slowly converting all of the hydrated crystals to the anhydrous brown form. After all of the crystals appear to be brown, continue heating gently, moving the burner back and forth judiciously, for an additional two minutes. Remove the burner, cover the crucible to minimize rehydration, and let it cool for about 15 minutes. Remove the cover, and slowly roll the brown crystals around the crucible. If some green crystals remain, repeat the heating process. Finally, weigh the cool uncovered crucible and its contents accurately.

Transfer the brown crystals in the crucible to an empty 50-mL beaker. Rinse out the crucible with two 5- to 7-mL portions of distilled water, and add the rinsings to the beaker. Swirl the beaker gently to dissolve the brown solid. The color will change to green as the copper ions are rehydrated. Measure out about 20 cm of 20-gauge aluminum wire (~0.25 g) and form the wire into a loose spiral coil. Put the coil into the solution so that it is completely immersed. Within a few moments you will observe some evolution of H_2, hydrogen gas, and the formation of copper metal on the Al wire. As the copper ions are reduced, the color of the solution will fade. The Al metal wire will be slowly oxidized and enter the solution as aluminum ions.

When the reaction is complete, which will take about 30 minutes, the solution will be colorless, and most of the copper metal that was produced will be on the Al wire. Add 5 drops of 6 M HCl to dissolve any insoluble aluminum salts and clear up the solution. Use your glass stirring rod to remove the copper from the wire as completely as you can. Slide the unreacted aluminum wire up the wall of the beaker with your stirring rod, and, while the wire is hanging from the rod, rinse off any remaining Cu particles with water from your wash bottle. If necessary, complete the removal of the Cu with a drop or two of 6 M HCl added directly to the wire. Put the wire aside; it has done its duty.

In the beaker you now have the metallic copper produced in the reaction, in a solution containing an aluminum salt. Set up a small Buchner funnel fitted with a moistened piece of filter paper. With light suction, decant the solution into the funnel. (Don't worry if some of the copper is also transferred.) Wash the copper metal in the beaker with small portions of distilled water; decant the wash into the funnel. Break up any copper particles with your stirring rod, and wash again, twice. Transfer the wash and the copper to the filter funnel. Wash any remaining copper into the funnel with water from your wash bottle. *All* of the copper must be transferred to the funnel. Rinse the copper on the paper once again with water. Turn off the suction. Add 10 mL of 95% ethanol to the funnel, and after a minute or so turn on the suction. Draw air through the funnel for about 5 minutes. Transfer the copper from the funnel to a weighed watch glass; this is perhaps most easily done by lifting the edge of the filter paper with a spatula and then lifting the paper from the funnel. The transfer must be quantitative; scrape any copper that adheres to the paper on to the watch glass with your spatula. Dry the copper on the watch glass under a heat lamp for 5 minutes. Allow to cool to room temperature and then weigh accurately.

Dispose of the liquid waste and copper produced in the experiment as directed by your instructor.

Data and Calculations: Determination of a Chemical Formula

Atomic masses: Copper _____   Cl _____   H _____   O _____

Mass of crucible   _____ g

Mass of crucible and hydrated sample   _____ g

Mass of hydrated sample   _____ g

Mass of crucible and dehydrated sample   _____ g

Mass of dehydrated sample   _____ g

Mass of empty watch glass   _____ g

Mass of watch glass and copper   _____ g

Mass of copper   _____ g

No. moles of copper   _____ moles

Mass of water evolved   _____ g

No. moles of water   _____ moles

Mass of chlorine in sample (by difference)   _____ g

No. moles of chlorine   _____ moles

Mole ratio, chlorine : copper in sample   _____ : 1

Mole ratio, water : copper in hydrated sample   _____ : 1

Formula of dehydrated sample (round to nearest integer)   _____

Formula of hydrated sample   _____

Advance Study Assignment: Determination of a Chemical Formula

1. To find the mass of a mole of an element, one looks up the atomic mass of the element in a table of atomic masses (see Appendix III or the Periodic Table). The molar mass of an element is simply the mass in grams of that element that is numerically equal to its atomic mass. For a compound substance, the molar mass is equal to the mass in grams that is numerically equal to the sum of the atomic masses in the formula of the substance. Find the molar mass of

    Cu _____ g   Cl _____ g   H _____ g   O _____ g H_2O _____ g

2. If one can find the ratio of the number of moles of the elements in a compound to one another, one can find the formula of the compound. In a certain compound of copper and oxygen, Cu_xO_y, we find that a sample weighing 0.5424 g contains 0.4831 g Cu.

    a. How many moles of Cu are there in the sample?

    $$\left(\text{No. moles} = \frac{\text{mass Cu}}{\text{molar mass Cu}} \right)$$

    _____ moles

    b. How many grams of O are there in the sample? (The mass of the sample equals the mass of Cu plus the mass of O.)

    _____ g

    c. How many moles of O are there in the sample?

    _____ moles

    d. What is the mole ratio (no. moles Cu/no. moles O) in the sample?

    _____ : 1

    e. What is the formula of the oxide? (The atom ratio equals the mole ratio, and is expressed using the smallest integers possible.)

    _____

    f. What is the molar mass of the copper oxide?

    _____ g

Experiment 5

IDENTIFICATION OF A COMPOUND BY MASS RELATIONSHIPS

In the previous experiment we showed how we can find the formula of a compound by analysis for the elements it contains. When chemical reactions occur, there is a relationship between the masses of the reactants and products that follows directly from the balanced equation for the reaction and the molar masses of the species that are involved. In this experiment we will use this relationship to identify an unknown substance.

Your unknown will be one of the following compounds, all of which are salts:

$$NaHCO_3 \quad Na_2CO_3 \quad KHCO_3 \quad K_2CO_3$$

In the first part of the experiment you will be heating a weighed sample of your compound in a crucible. If your sample is a carbonate, there will be no chemical reaction that occurs, but any small amount of adsorbed water will be driven off. If your sample is a hydrogen carbonate, it will decompose by the following reaction, using $NaHCO_3$ as the example:

$$2\,NaHCO_3(s) \rightarrow Na_2CO_3(s) + H_2O(g) + CO_2(g) \tag{1}$$

In this case there will an appreciable decrease in mass, since some of the products will be driven off as gases. If such a mass decrease occurs, you can be sure that your sample is a hydrogen carbonate.

In the second part of the experiment, we will treat the solid carbonate in the crucible with HCl, hydrochloric acid. There will be considerable effervescence as CO_2 gas is evolved; the reaction that occurs is, using Na_2CO_3 as our example:

$$Na_2CO_3(s) + 2\,H^+(aq) + 2\,Cl^-(aq) \rightarrow 2\,NaCl(s) + H_2O(l) + CO_2(g) \tag{2}$$

(Since HCl in solution exists as ions, we write the equation in terms of ions.) We then heat the crucible strongly to drive off any excess HCl and any water that is present, obtaining pure, dry, solid NaCl as our product.

To identify your unknown, you will need to find the molar masses of the possible reactants and final products. For each of the possible unknowns there will be a different relationship between the mass of the original sample and the mass of the chloride salt that is produced in Reaction 2. If you know your sample is a carbonate, you need only be concerned with the mass relationships in Equation 2, and should use as the original mass of your unknown the mass of the carbonate after it has been heated. If you have a hydrogen carbonate, the overall reaction your sample undergoes will be the sum of Reactions 1 and 2. From your experimental data you will be able to calculate the ratio of the mass of the solid chloride to the mass of either the original hydrogen carbonate or the mass of the anhydrous carbonate in your sample of unknown. From your calculation of the relative masses of solid chloride to solid hydrogen carbonate or solid carbonate in Equations 1 and 2 you can calculate what the theoretical ratio of those masses should be. Your observed value should match one of the theoretical values and thus allow you to identify the compound your unknown contains.

Experimental Procedure

Obtain an unknown from the stockroom.

Clean your crucible and its cover by rinsing them with distilled water and then drying them with a towel. Place the crucible with its cover slightly ajar on a clay triangle. Heat gently with your bunsen burner flame for a minute or two and then strongly for two more minutes. Allow the crucible and cover to cool to room temperature (it will take about 10 minutes).

Weigh the crucible and cover accurately on an analytical balance. Record the mass on the Data page. With a spatula, transfer about 0.5 g of the unknown to the crucible. Weigh the crucible, cover, and the sample of unknown on the balance. Record the mass.

Put the crucible on the clay triangle, with the cover ajar. Heat the crucible, gently and intermittently, for a few minutes. Gradually increase the flame intensity, to the point where the bottom of the crucible is at red heat. Heat for 10 minutes. Allow the crucible to cool for 10 minutes, and then weigh it, with its cover and contents on the analytical balance, recording the mass.

At this point the sample in the crucible is a dry carbonate, since the heating process will convert any hydrogen carbonate to carbonate.

Put the crucible on the clay triangle, leaving the cover off. Add about 25 drops of 6 M HCl, a drop at a time, to the sample. As you add each drop, you will probably observe effervescence as CO_2 is produced. Let the action subside before adding the next drop, to keep the effervescence confined to the lower part of the crucible. We do not want the product to foam up over the edge. When you have added all of the HCl, the effervescence should have ceased, and the solid should be completely dissolved. Heat the crucible gently for brief periods to complete the solution process. If all of the solid is not dissolved, add 6 more drops of 6 M HCl and warm gently.

Place the cover on the crucible in an off-center position, to allow water to escape during the next heating operation. Heat the crucible, gently and intermittently, for about 10 minutes, to slowly evaporate the water and excess HCl. If you heat too strongly, spattering will occur and you may lose some sample. When the sample is dry, gradually increase the flame intensity, finally getting the bottom of the crucible to red heat. Heat at full flame strength for 10 minutes.

Allow the crucible to cool for 10 minutes. Weigh it, with its cover and contents, on the analytical balance, recording the mass.

Dispose of the sample by dissolving it in water and pouring it down the drain.

Data and Calculations: Identification of a Compound by Mass Relationships

Atomic masses: Na _____ K _____ H _____ C _____ O _____ Cl _____

Molar masses: $NaHCO_3$ _____ g Na_2CO_3 _____ g NaCl _____ g

                $KHCO_3$ _____ g K_2CO_3 _____ g KCl _____ g

Mass of crucible and cover plus unknown   _____ g

Mass of crucible and cover   _____ g

Mass of unknown   _____ g

Mass of crucible, cover, and unknown after heating   _____ g

Loss of mass of sample   _____ g

Sample is a   carbonate   hydrogen carbonate   (underline your choice)

Mass of crucible, cover, and solid chloride   _____ g

Mass of solid chloride   _____ g

Ratio   mass of chloride : mass of dry carbonate
           (if your sample is a carbonate)   _____ : 1
           or
           mass of chloride : mass of hydrogen carbonate
           (if your sample is a hydrogen carbonate)   _____ : 1

Theoretical ratios

    If sample is a carbonate, use Equation 2

          1 mole Na_2CO_3 → _____ moles NaCl

          _____ g Na_2CO_3 → _____ g NaCl   Mass ratio _____ : 1
                                               (NaCl to Na_2CO_3)

          1 mole K_2CO_3 → _____ moles KCl

          _____ g K_2CO_3 → _____ g KCl   Mass ratio _____ : 1
                                               (KCl to K_2CO_3)

(continued on following page)

(continued)

If sample is a hydrogen carbonate, use Equation 1 plus Equation 2

1 mole $NaHCO_3$ → _____ moles NaCl

_____ g $NaHCO_3$ → _____ g NaCl   Mass ratio _____ : 1
     (NaCl to $NaHCO_3$)

1 mole $KHCO_3$ → _____ moles KCl

_____ g $KHCO_3$ → _____ g KCl   Mass ratio _____ : 1
     (KCl to $KHCO_3$)

Identity of unknown _____   Unknown No. _____

Advance Study Assignment: Identification of a Compound by Mass Relationships

1. A student attempts to identify an unknown compound by the method used in this experiment. She finds that when she heated a sample weighing 0.5015 g the mass went down appreciably, to 0.3432 g. When the product was converted to a chloride, the mass went up, to 0.3726 g.

   a. Is the sample a carbonate? _____ Give your reasoning.

   b. What are the two compounds that might be in the unknown?

   _____ and _____

   c. Write the chemical equation for the overall reaction that would occur when the original compound was converted to a chloride. If the compound is a hydrogen carbonate, use the sum of Equations 1 and 2. If the sample is a carbonate, use Equation 2. Write the equation for a sodium salt and then for a potassium salt.

   d. How many moles of the chloride salt would be produced from one mole of original compound?

   _____

   e. How many grams of the chloride salt would be produced from one molar mass of original compound?

   Molar masses: $NaHCO_3$ _____ g Na_2CO_3 _____ g $NaCl$ _____ g

   $KHCO_3$ _____ g K_2CO_3 _____ g KCl _____ g

   If a sodium salt,   _____ g original compound → _____ g chloride

   If a potassium salt,   _____ g original compound → _____ g chloride

   f. What is the theoretical mass ratio, mass chloride/mass original compound?

   If she has an Na salt, _____ : 1   If she has a K salt, _____ : 1

   (continued)

(continued)

g. What was the observed mass ratio, mass chloride/mass original salt?

____________ : 1

h. Which compound did the student have as an unknown?

____

xperiment 6

ANALYSIS OF AN UNKNOWN CHLORIDE

One of the important applications of precipitation reactions lies in the area of quantitative analysis. Many substances that can be precipitated from solution are so slightly soluble that the precipitation reaction by which they are formed can be considered to proceed to completion. Silver chloride is an example of such a substance. If a solution containing Ag^+ ion is slowly added to one containing Cl^- ion, the ions will react to form AgCl:

$$Ag^+(aq) + Cl^-(aq) \rightarrow AgCl(s) \tag{1}$$

Silver chloride is so insoluble that essentially all of the Ag^+ added will precipitate as AgCl until all of the Cl^- is used up. When the amount of Ag^+ added to the solution is equal to the amount of Cl^- initially present, the precipitation of Cl^- ion will be, for all practical purposes, complete.*

A convenient method for chloride analysis using AgCl has been devised. A solution of $AgNO_3$ is added to a chloride solution just to the point where the number of moles of Ag^+ added is equal to the number of moles of Cl^- initially present. We analyze for Cl^- by simply measuring how many moles of $AgNO_3$ are required. Surprisingly enough, this measurement is rather easily made by an experimental procedure called a titration.

In the titration a solution of $AgNO_3$ of known concentration (in moles $AgNO_3$ per liter of solution) is added from a calibrated buret to a solution containing a measured amount of unknown. The titration is stopped when a color change occurs in the solution, indicating that stoichiometrically equivalent amounts of Ag^+ and Cl^- are present. The color change is caused by a chemical reagent, called an indicator, that is added to the solution at the beginning of the titration.

The volume of $AgNO_3$ solution that has been added up to the time of the color change can be measured accurately with the buret, and the number of moles of Ag^+ added can be calculated from the known concentration of the solution.

In the Mohr method for the volumetric analysis of chloride, which we will employ in this experiment, the indicator used is K_2CrO_4. The chromate ion present in solutions of this substance will react with silver ion to form a red precipitate of Ag_2CrO_4. Under the conditions of the titration, the Ag^+ added to the solution reacts preferentially with Cl^- until that ion is essentially quantitatively removed from the system, at which point Ag_2CrO_4 begins to precipitate and the solution color changes from yellow to buff. The end point of the titration is that point at which the color change is first observed.

In this experiment, weighed samples containing an unknown percentage of chloride will be titrated with a standardized solution of $AgNO_3$, and the volumes of $AgNO_3$ solution required to reach the end point of each titration will be measured. Given the molarity of the $AgNO_3$

$$\text{no. of moles } Ag^+ = \text{no. of moles } AgNO_3 = M_{AgNO_3} \times V_{AgNO_3} \tag{2}$$

where the volume of $AgNO_3$ is expressed in liters and the molarity M_{AgNO_3} is in moles per liter of solution. At the end point of the titration,

---

* See Experiment 24 for a discussion of the principles governing precipitations of this sort.

$$\text{no. of moles } Ag^+ \text{ added} = \text{no. of moles } Cl^- \text{ present in unknown} \qquad (3)$$

$$\text{no. of grams } Cl^- \text{ present} = \text{no. of moles } Cl^- \text{ present} \times \text{MM Cl} \qquad (4)$$

$$\% \text{ Cl} = \frac{\text{no. of grams } Cl^-}{\text{no. of grams unknown}} \times 100 \qquad (5)$$

Experimental Procedure

Obtain from the stockroom a buret and a vial containing a sample of an unknown solid chloride. Weigh out accurately on the analytical balance three samples of the chloride, each sample weighing about 0.2 grams. This weighing is best done by accurately weighing the vial and its contents and pouring out the sample a little at a time into a 250-mL Erlenmeyer flask until the vial has lost about 0.2 g of chloride sample. Again weigh the sample vial accurately to obtain the exact amount of chloride sample poured into the flask. Put two other samples of similar mass into clean, dry, small beakers, weighing the vial accurately after the size of each sample has been decided upon. Add 50 mL of distilled water to the flask to dissolve the sample and add 3 drops of some 1 M K_2CrO_4 indicator solution. Using the graduated cylinder at the reagent shelf, measure out about 100 mL of the standardized $AgNO_3$ solution into a clean *dry* 125-mL Erlenmeyer flask. This will be your total supply for the entire experiment so do not waste it. Clean your buret thoroughly with soap solution and rinse it with distilled water. Pour three successive 2- or 3-mL portions of the $AgNO_3$ solution into the buret and tip it back and forth to rinse the inside walls. Allow the $AgNO_3$ solution to drain out the buret tip completely each time. Fill the buret with the $AgNO_3$ solution. Open the buret stopcock momentarily to flush any air bubbles out of the tip of the buret. Be sure your stopcock fits snugly and that the buret does not leak. (See Appendix IV for procedures regarding titrations with a buret.)

Read the initial buret level to 0.02 mL. You may find it useful when making readings to put a white card marked with a thick black stripe behind the meniscus. If the black line is held just below the level to be read, its reflection in the surface of the meniscus will help you obtain an accurate reading. Begin to add the $AgNO_3$ solution to the chloride solution in the Erlenmeyer flask. A white precipitate of AgCl will form immediately, and the amount will increase during the course of the titration. At the beginning of the titration, you can add the $AgNO_3$ fairly rapidly, a few milliliters at a time, swirling the flask as best you can to mix the solution. You will find that at the point where the $AgNO_3$ hits the solution, there will be a red spot of Ag_2CrO_4, which disappears when you stop adding nitrate and swirl the flask. As you proceed with the titration, the red spot will persist more and more, since the amount of excess chloride ion, which reacts with the Ag_2CrO_4 to form AgCl, will slowly decrease. Gradually decrease the rate at which you add $AgNO_3$ as the red color becomes stronger. At some stage you may find it convenient to set your buret stopcock to deliver $AgNO_3$ slowly, drop by drop, while you swirl the flask. When you are near the end point, add the $AgNO_3$ drop by drop, swirling between drops. The end point of the titration is that point where the mixture first takes on a permanent buff or reddish-yellow color that does not revert to pure yellow on swirling. If you are careful, you can hit the end point within one drop of $AgNO_3$. When you have reached the end point, stop the titration and record the buret level.

Pour the solution you have just titrated into another 125-mL Erlenmeyer flask or into a 250-mL beaker. To that solution add a few milliliters of 0.1 M NaCl; the color of the mixture should go back to the original yellow. Use the color of this mixture as a reference against which you compare your samples in the remaining titrations.

Rinse out the 250-mL Erlenmeyer flask in which you carried out the titration. Take your second sample and carefully pour it from the beaker into the Erlenmeyer flask. Wash out the beaker a few times with distilled water from your wash bottle and pour the washings into the flask. *All of the sample* must be transferred if the analysis is to be accurate. Add water to the flask to a volume of about 50 mL and swirl to dissolve the solid. Refill your buret, take a volume reading, add the indicator, and proceed to titrate to an end point as before. This titration should be more accurate than the first, since the volume of $AgNO_3$ used is proportional to sample size and therefore can be estimated rather well on the basis of the relative masses of the two samples. In addition, you have a reference for color comparison that should make it easier to recognize when a color change has occurred.

Titrate the third sample as you did the second. With care it should be possible to obtain volume-mass ratios that agree to within less than 1% in the last two titrations.

Silver nitrate is very expensive. Pour all titrated solutions and any $AgNO_3$ remaining in your buret or flask into the waste bottles provided unless directed otherwise by your instructor.

Data and Calculations: Analysis of an Unknown Chloride

Molarity of standard $AgNO_3$ solution _____ M

	I	*II*	*III*
Mass of vial and chloride unknown	_____ g	_____ g	_____ g
Mass of vial less sample	_____ g	_____ g	_____ g
Initial buret reading	_____ mL	_____ mL	_____ mL
Final buret reading	_____ mL	_____ mL	_____ mL
Mass of sample	_____ g	_____ g	_____ g
Volume of $AgNO_3$ used to titrate sample	_____ mL	_____ mL	_____ mL
No. of moles of $AgNO_3$ used to titrate sample	_____	_____	_____
No. of moles of Cl^- present in sample	_____	_____	_____
Mass of Cl^- present in sample	_____ g	_____ g	_____ g
Percentage of Cl^- in sample	_____ %	_____ %	_____ %

Mean value of percentage of Cl^- in unknown _____ %

Unknown no. _____

Advance Study Assignment: Determination of an Unknown Chloride

1. A sample containing 0.187 g Cl^- is dissolved in 50.0 mL water.

   a. How many moles of Cl^- ion are in the solution?

   _____ moles Cl^-

   b. What is the molarity of the Cl^- ion in the solution? ($M_{Cl^-} = n_{Cl^-}/V_{soln}$)

   _____ M

2. A solid chloride sample weighing 0.2585 g required 47.28 mL of 0.05689 M $AgNO_3$ to reach the Ag_2CrO_4 end point.

   a. How many moles Cl^- ion were present in the sample? (Use Eqs. 2 and 3.)

   _____ moles Cl^-

   b. How many grams Cl^- ion were present? (Use Eq. 4.)

   _____ g Cl^-

   c. What was the mass percent Cl^- ion in the sample? (Use Eq. 5.)

   _____ % Cl^-

3. How would the following errors affect the mass percent Cl^- obtained in Question 2c? Give your reasoning in each case.

   a. The student read the molarity of $AgNO_3$ as 0.05896 M instead of 0.05689 M.

   b. The student was past the end point of the titration when he took the final buret reading.

Experiment 7

DETERMINATION OF THE BAROMETRIC PRESSURE—CHARLES' LAW

Gases differ from liquids and solids in that much of the ordinary physical behavior of gases can be described by simple laws that apply to all gases. Few such relations exist for liquids and solids, so information about their physical properties must be obtained by direct experiment on the liquid or solid of interest.

Gases are easily compressed relative to liquids and solids. If one doubles the pressure on a gas, the volume decreases by a factor of two. Over rather wide ranges of pressure, the pressure-volume product, *PV*, remains nearly constant as long as the temperature remains constant. This relationship is called Boyle's Law, after Robert Boyle, who discovered it in 1660. It was one of the first natural laws that scientists recognized.

When a gas is heated at constant volume, the pressure goes up. In 1787, about 120 years after Boyle did his work, Charles found that the pressure increased linearly with Celsius temperature. Boyle was in no position to discover Charles' Law, since in 1660 the idea of temperature was not well developed. In 1848, Lord Kelvin, at the ripe old age of 24, recognized that one could set up an absolute temperature scale on which the pressure was actually proportional to temperature. This idea was relatively sophisticated and led to the observation that many natural laws, not just the gas laws, are most simply expressed on the absolute temperature scale, now called the Kelvin scale.

In this experiment we will first determine the barometric pressure in the laboratory. To do this we will use a method suggested by an experiment which you have probably done at home. If you take an inverted drinking glass and immerse it in water, you find that as you push the glass down, the air in the glass is slowly compressed. You have to immerse the glass to a sizable depth to decrease the volume of air significantly. It turns out that if you took it to a depth of about 10 meters, you would find that the volume would decrease to about half its value at the surface. At that point the pressure on the air would be about two atmospheres. This means that 10 meters of water add about one atmosphere of pressure, or, putting it another way, the barometric pressure is about equal to the pressure exerted by about 10 meters of water, or about 750 mm Hg, or about 1×10^5 Pascals, or about 15 lb/in². While this experiment might be interesting to perform (you could do it if you are a SCUBA diver), it is a bit tricky to do in the laboratory. So we will resort to a piece of apparatus that uses the same principle but allows easier measurement of volume and pressure changes.

In the Charles' Law experiment, we will heat a sample of air of fixed mass, holding the volume essentially constant while we measure the increase in pressure. The rate of change in pressure with temperature will illustrate Charles' Law and also allow us to set up the absolute temperature scale.

Experimental Procedure

WEAR YOUR SAFETY GLASSES WHILE PERFORMING THIS EXPERIMENT

Obtain from the stockroom a 50-mL buret, a 500-mL flask, a meter stick, a piece of glass tubing about 60 cm long and 25 mm in diameter, a U-tube mercury manometer fitted with a piece of rubber tubing, and a plastic bucket.

Determining the Barometric Pressure

Success in this experiment requires paying attention to several factors. The temperature of the entire apparatus must remain constant and equal to the temperature in the laboratory. The apparatus must be air-tight, with no leaks at the stopcock or at the flask.

Measure the temperature of the air in the laboratory with your thermometer. If the thermometer has been in your drawer, it may take a minute or two for it to come to room temperature. When the reading gets steady, record the temperature on the Data page. Add water from the cold tap at the sink to your 600-mL beaker, filling it about 2/3 full. Measure the temperature of the water. With stirring, add small amounts of warm or cold water from the tap as necessary to bring the water in the beaker to as close to room temperature as you can. Stir for a minute or so and read the temperature again, to make sure you got it right. Put a solid rubber stopper firmly in the end of the piece of glass tubing, and support the tube vertically on the lab bench with a clamp. Pour water from the beaker into the tube, filling it about 2/3 full (see Fig. 7.1).

With the stopcock closed, pour water from the beaker into the buret until it is about full. Drain the water into the sink. If the buret is clean, it will drain smoothly, leaving no droplets on the wall. If it is clean, proceed to the next paragraph. If droplets appear, clean the buret with detergent and warm water from the tap, using a buret brush. Drain the buret, and rinse with cold tap water until no suds remain. Then rinse it twice more with water from the beaker.

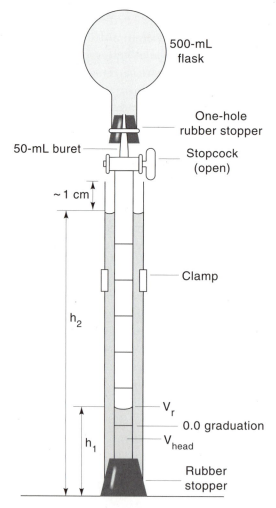

FIGURE 7.1

With the buret inverted, open the stopcock, and blow any water *down* out of the tip and stopcock. Insert the small end of the buret into a one-hole stopper. Moisten the stopper and attach it firmly to the 500-mL flask, as in Figure 7.1. If the stopcock is plastic, tighten the nut holding the stopcock in place so as to prevent any leakage.

Slowly immerse the buret in the water, lowering it until the end touches the rubber stopper. As you lower the buret, the pressure of the air inside will gradually increase; the volume of the air will go down as water rises in the buret. Wait for a minute or two until the water level inside the buret gets steady. If the water level continues to rise, you have a leak in the system, most likely at the stopcock. Loosen and remove the stopcock. Spread a drop of glycerine over the surface, reinsert the stopcock and tighten the nut that holds it. Lift the buret out of the tube, let it drain for a minute or so, and then lower it as you did before.

When the level is steady, add some water from the beaker to the long tube until the water level is about 1 cm below the top. You are then ready to make pressure and volume measurements. The increase in pressure, ΔP, of the air in the apparatus is equal, in mm H_2O, to the difference in height of the water level in the tube and that of the water level in the buret. Measure the height to ± 1 mm of each level above the lab bench, using your meter stick. Record those two levels.

The change in volume, ΔV, of the air will equal the volume reading of the water level in the buret, V_r, plus the volume, V_{head}, that lies beyond the 0.0 graduation on the buret. Read and record the volume reading on the buret to ± 0.1 mL. Take account of the fact that the buret is upside down, so readings will increase as you go up the buret. (See Appendix IV for a discussion of burets.)

Raise the buret about 1/4 of its length, hold it in place with a clamp, and make readings of heights and volume V_r as before.

Lift the buret out of the tube, let it drain for a minute or so, and repeat the experiment. The heights and volume reading should be about the same (± 1 mm or 0.1 mL) as in your first trial. If they are not, make one or more further trials, until you get consistent readings.

To calculate the barometric pressure, we make the assumption that Boyle's Law holds for the sample of air in the system. If we take P_2 and V_2 to represent the pressure and volume in the final state, and P_1 and V_1 for the initial state, we have, by Boyle's Law in its familiar form,

$$P_2 V_2 = P_1 V_1 \tag{1}$$

The initial pressure, P_1, taken when the lower end of the buret is open to the air in the laboratory, is equal to the barometric pressure, P_{bar}. The final pressure is equal to P_{bar} plus the change in pressure ΔP due to the head of water, $h_2 - h_1$, caused by the immersion of the buret. If we express all pressures in mm of H_2O, then

$$P_2 = P_1 + \Delta P = (P_{bar} + h_2 - h_1)\text{mm } H_2O \quad \text{and} \quad P_1 = P_{bar} \tag{2}$$

The volumes V_2 and V_1 are a bit harder to find. V_1 is the volume of the air in the system at P_1 and is equal to the volume of the flask plus the total volume of the buret. V_2 is equal to V_1 minus the change in volume ΔV that occurs when the buret is put into the water.

We measure those volumes by first finding the mass of water that will fill the flask and the buret. Detach the flask from the buret, noting the location of the top of the stopper in the neck of the flask. Fill the flask with water to that level. Pour the water into a weighed 600-mL beaker and reweigh to 0.1 g on a top-loading balance. The volume V_{flask} of the flask in milliliters is just about equal to the mass of the water in grams, since the density of water is just about 1.00 g/mL. (If there is no balance available with enough capacity, use a 1000-mL graduated cylinder to measure the volume.) Then fill the buret completely with water, including the tip. Close the stopcock to complete the filling. Drain the buret completely into a weighed 100-mL beaker and weigh to ± 0.1 g. Calculate the volume of the buret, V_{buret}, again assuming the density of water to be 1 g/mL. The initial volume V_1 of the system equals $V_{flask} + V_{buret}$.

The final volume V_2 of the air in the system is equal to V_1, which we now know, minus the change in the volume ΔV that occurs when the buret is immersed. The magnitude of ΔV is equal to the volume of the buret, V_{head}, that lies above the 0.0 graduation plus the reading V_r of the water level in the buret. To find V_{head}, fill the buret with water, right to the top. Open the stopcock and let the water run into the weighed 100-mL beaker until you get to the 0.0 level. Reweigh the beaker and water, and get the value of V_{head} from the mass of the water as you did with the other measured volumes. It turns out that, because of the way we make our volume measurements, ΔV will be about 0.2 mL too small, so we will add that volume to ΔV when making the calculations of barometric pressure. ΔV must be subtracted from the initial volume, since the volume of the air goes down when the buret is put into the water.

We can now proceed with the calculation of P_{bar}. Recalling Boyle's Law,

$$P_2 V_2 = P_1 V_1 \tag{1}$$

$$P_2 = P_1 + \Delta P \quad \text{and} \quad V_2 = V_1 - \Delta V \qquad P_1 = P_{bar} \quad \text{and} \quad V_1 = \text{initial volume}$$

So we have

$$(P_1 + \Delta P)(V_1 - \Delta V) = P_1 V_1$$

$$P_1 V_1 + V_1 \Delta P - P_1 \Delta V - \Delta V \Delta P = P_1 V_1$$

Solving for P_1,

$$P_1 = \frac{V_1 \Delta P - \Delta V \Delta P}{\Delta V} = P_{bar} \tag{3}$$

where

$$V_1 = (V_{flask} + V_{buret}) \quad \text{and} \quad \Delta V = (V_{head} + V_r + 0.2) \text{ mL} \tag{4}$$

Make the calculations of P_{bar}, using Equations 2, 3, and 4, for each of the buret heights you used in this experiment. Convert the barometric pressure in mm H_2O to mm Hg by multiplying by the ratio of the density of water to the density of mercury, which is 1/13.57. Compare your values with that obtained with the barometer in the laboratory.

Charles' Law

In this part of the experiment we will measure the pressure exerted by a sample of dry air as it is heated at constant volume.

We will be setting up the apparatus shown in Figure 7.2. The gas will be contained in a dry 250-mL Erlenmeyer flask. If the flask is wet, shake out any residual water and rinse it well with a few milliliters of acetone. Pour out the acetone, and gently, gently blow compressed air into the flask for a minute or two. (Keep the noise down; we don't want the lab to sound like an airport at rush hour.) When the flask is *completely dry,* firmly insert the stopper after moistening it, and clamp the flask in a 1000- or 600-mL beaker as shown. Using a split stopper, clamp the mercury manometer so that it is near the flask and over the plastic ice cream bucket. (**CAUTION:** *Mercury vapor is hazardous, and the liquid, which was called quicksilver in the early days, is hard to clean up. Use due care in handling the manometer.)*

Attach the piece of rubber tubing to the flask. If it is not already attached to the manometer, carefully connect it. You should now have a fixed amount of dry air in the flask and tubing, at a pressure very close to barometric. Fill a 400-mL beaker with ice-water from the

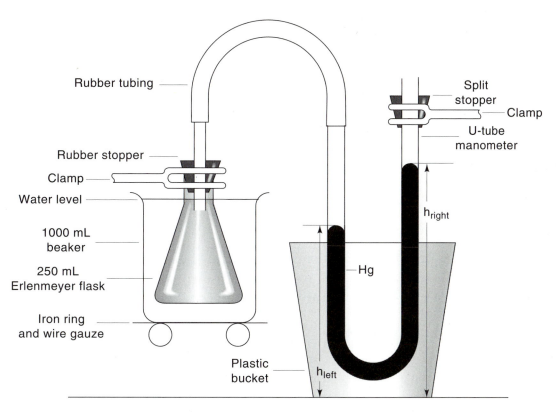

FIGURE 7.2

supply bucket and pour it into the 1000-mL beaker. Refill the beaker with ice-water and use that to fill the larger beaker up to within about 1 cm of the top. Stir the cold water for two minutes, and then measure the temperature as precisely as your thermometer allows. The temperature should be somewhere near 5°C. Using a ruler or meter stick measure the heights, in millimeters above the lab bench, of the mercury levels in the left and right arms of the manometer. Record the temperature and the two heights.

Heat the water in the beaker with your bunsen burner until the temperature is about 20°C. Remove heat, and stir for a few minutes to bring all the water to the same temperature. When the temperature is steady, read the Hg levels in the manometer and record your values of temperature and height.

Make two more sets of readings, one at about 40°C and the other at about 60°C. (**If the mercury level gets to within 2 cm of the top of the manometer or within 1 cm of the beginning of the bend at the bottom of the U-tube, stop heating!** This should not happen, but if your U-tube is too short, it could, so pay attention as you heat the water.) At these higher temperatures, the water bath will tend to cool fairly quickly if no heat is applied. So, when you are at about the right temperature, remove the burner and stir the water. The temperature will tend to go up for a while, since the iron ring is hot, and then will level off and start to go down. When it is steady, make the temperature and manometer readings. Do not disassemble the apparatus until the water in the large beaker has cooled to the point where it is comfortable to touch.

Return the borrowed equipment to the stockroom after you have completed the calculations based on your data.

Data and Calculations: Determining the Barometric Pressure and Charles' Law Barometric Pressure

Trial 1	Buret position	Height, h_2, of water in tube	Height, h_1, of water in buret	Volume reading, V_r, in buret
A	fully immersed	_____ mm	_____ mm	_____ mL
B	3/4 immersed	_____ mm	_____ mm	_____ mL

Trial 2

A	fully immersed	_____ mm	_____ mm	_____ mL
B	3/4 immersed	_____ mm	_____ mm	_____ mL

Mass of 600-mL beaker _____ g   Temperature in the lab _____ °C

Mass of beaker plus
water in flask _____ g   Volume of flask, V_{flask} _____ mL

Mass of 100-mL beaker _____ g   (Density of water $\cong$ 1.00 g/mL)

Mass of beaker plus
water in full buret _____ g   Volume of buret, V_{buret} _____ mL

Mass of beaker plus
water above 0.0 level _____ g   Volume of head, V_{head} _____ mL

$V_1 = V_{flask} + V_{buret}$ _____ mL

$$\frac{P \text{ in mm Hg}}{P \text{ in mm H}_2\text{O}} = \frac{1}{13.57} = \frac{d_{H_2O}}{d_{Hg}}$$

Trial 1	$\Delta P = h_2 - h_1$	$\Delta V = V_{head} + V_r$ + 0.2 mL	P_{bar} (by Equation 3)
A	_____ mm H$_2$O	_____ mL	_____ mm H$_2$O or _____ mm Hg
B	_____ mm H$_2$O	_____ mL	_____ mm H$_2$O or _____ mm Hg

(continued on following page)

(continued)

Trial 2	ΔP	ΔV	P_{bar}
A	_____ mm H_2O	_____ mL	_____ mm H_2O or _____ mm Hg
B	_____ mm H_2O	_____ mL	_____ mm H_2O or _____ mm Hg

P_{bar} from laboratory barometer _____ mm Hg

(Make your calculations of P_{bar} below)

Charles' Law

t, °C	*Height of level in left arm*	*Height of level in right arm*	$\Delta h, h_{rt} - h_{left}$	$P = P_{bar} + \Delta h$
_____	_____ mm	_____ mm	_____ mm	_____ mm Hg
_____	_____ mm	_____ mm	_____ mm	_____ mm Hg
_____	_____ mm	_____ mm	_____ mm	_____ mm Hg
_____	_____ mm	_____ mm	_____ mm	_____ mm Hg

On the graph paper at the end of this section, plot the pressure P as a function of the temperature in °C. (See Appendix V for a discussion of graphing techniques.) To make the graph, enter each pressure and temperature on the grid of lines at the point associated with that pair of values. Make your entries for the four pairs of data points you obtained in the experiment. Put the best straight line you can through the points (minimize the sum of the distances from each point to the line). If you have access to Cricket Graph on a computer, you may make the graph using that software, which will also find the best straight line through the points, along with its equation. Comment below on the degree to which your points fall on a straight line.

We can use the graph to set up the Kelvin absolute temperature scale. On that scale pressure is proportional to temperature, for a sample of gas heated at constant volume; for a gas heated between states 1 and 2,

$$\frac{P_2}{P_1} = \frac{T_2}{T_1} = \frac{t_2 + A}{t_1 + A} \qquad (5)$$

where T_2 and T_1 are Kelvin temperatures. A is a constant, since the Celsius degree is equal in size to a Kelvin degree.

(continued on following page)

(continued)

To find A, select two temperatures, t_2 and t_1, one near the right end of your graph and one near the left. Find the pressures, P_2 and P_1, that lie on the line at those temperatures:

t_2 ——————— °C t_1 ——————— °C P_2 ——————— mm Hg P_1 ——————— mm Hg

Set up Equation 5, using these values:

Solve the equation for the value of A.

$$A = \text{_____} \ K$$

Evaluate T_2 and T_1, using Equation 5:

$$T_2 = \text{_____} \ K; \ T_1 = \text{_____} \ K$$

On the basis of your data, Charles' Law is obeyed, as you can show by substituting into Equation 5. Your value of A is probably not equal to 273, which is obtained by a more accurate experiment, but it should be fairly close. Can you suggest why your value of A is likely to be larger than the literature value?

If you have used Cricket Graph, you will get an equation for P vs t (y vs x if you didn't change the names of the variables). From that equation it is easy to find the value of t when P equals zero. Then, looking at Equation 5, it is clear that P_2 will equal zero when t_2 equals $-A$. (P_1 can be anything, as long as it is not zero; under such conditions, $t_1 + A$ will not equal zero.) Find the value of t_2 when P_2 is 0, and from that the value of A. Your result should be just about the same as that obtained if you used the graph paper.

$$t_2 = \text{_____} \ °C; \ A = \text{_____}$$

Name _____ Section _____

Charles' Law
(Data and Calculations)

Advance Study Assignment: Determination of Barometric Pressure and Charles' Law

1. The barometric pressure on a summer day in Minnesota was found to be 743 mm Hg.

   a. What would that pressure be in mm H_2O? (See DATA page.)

   _____ mm H_2O

   b. How many meters of water would it take to exert that barometric pressure?

   _____ m H_2O

   c. Why can't you lift water more than about 10 meters from a well, using a suction pump?

2. In a Charles' Law experiment like this one, a student finds that the Hg level in the right arm of the manometer in Figure 7.2 is 15.2 cm higher than that in the left arm.

   a. Is the pressure in the flask greater or less than barometric pressure?

   (underline your answer)   greater   less

   b. If the barometric pressure is that in Problem 1, what is the total pressure in the flask?

   _____ mm Hg

3. When a student heated a sample of dry gas from 20°C to 30°C in the apparatus in Figure 7.2, she found that the total pressure increased from 740 to 765 mm Hg. Use these data to find the Kelvin temperature at 20°C; at 30°C. (Use Eq. 5.)

   _____ K; _____ K

Experiment 8

MOLAR MASS OF A VOLATILE LIQUID

One of the important applications of the Ideal Gas Law is found in the experimental determination of the molar masses of gases and vapors. In order to measure the molar mass of a gas or vapor we need simply to determine the mass of a given sample of the gas under known conditions of temperature and pressure. If the gas obeys the Ideal Gas Law,

$$PV = nRT \tag{1}$$

If the pressure P is in atmospheres, the volume V in liters, the temperature T in K, and the amount n in moles, then the gas constant R is equal to 0.0821 L atm/(mole K).

From the measured values of P, V, and T for a sample of gas we can use Equation 1 to find the number of moles of gas in the sample. The molar mass in grams, MM, is equal to the mass g of the gas sample divided by the number of moles n.

$$n = \frac{PV}{RT} \qquad MM = \frac{g}{n} \tag{2}$$

This experiment involves measuring the molar mass of a volatile liquid by using Equation 2. A small amount of the liquid is introduced into a weighed flask. The flask is then placed in boiling water, where the liquid will vaporize completely, driving out the air and filling the flask with vapor at barometric pressure and the temperature of the boiling water. If we cool the flask so that the vapor condenses, we can measure the mass of the vapor and calculate a value for MM.

Experimental Procedure*

Obtain a special round-bottomed flask, a stopper and cap, and an unknown liquid from the storeroom. Support the flask on an evaporating dish or in a beaker at all times. If you should break or crack the flask, report it to your instructor immediately so that it can be repaired. With the stopper loosely inserted in the neck of the flask, weigh the empty, dry flask on the analytical balance. Use a copper loop, if necessary, to suspend the flask from the hook supporting the balance pan. With an automatic balance, use a cork ring to support the flask.

Pour about half your unknown liquid, about 5 mL, into the flask. Assemble the apparatus as shown in Figure 8.1. Place the cap on the neck of the flask. Add a few boiling chips to the water in the 600-mL beaker and heat the water to the boiling point. Watch the liquid level in your flask; the level should gradually drop as vapor escapes through the cap. After all the liquid has disappeared and no more vapor comes out of the cap, continue to boil the water gently for 5 to 8 minutes. Measure the temperature of the boiling water. Shut off the burner and wait until the water has stopped boiling (about 1/2 minute) and then loosen the clamp holding the flask in place. Slide out the flask, remove the cap, and *immediately* insert the stopper used previously.

---

* See W. L. Masterton and T. R. Williams, J. Chem. Educ. *36*, 528 (1959). For an alternate apparatus, see instructor's manual.

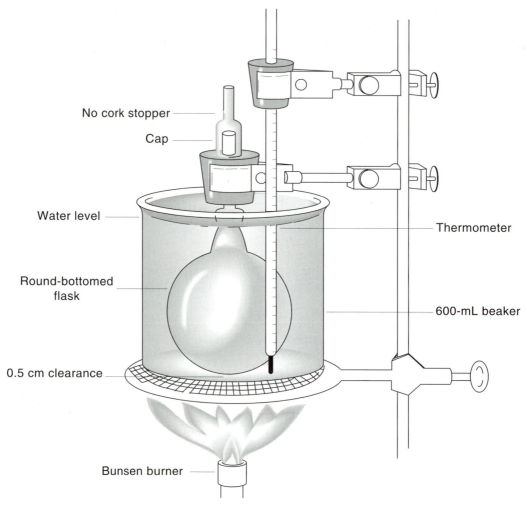

No cork stopper

Cap

Water level

Round-bottomed
flask

0.5 cm clearance

Bunsen burner

Thermometer

600-mL beaker

FIGURE 8.1

Remove the flask from the beaker of water, holding it by the neck, which will be cooler. Immerse the flask in a beaker of cool water to a depth of about 5 cm. After holding the flask in the water for about 2 minutes to allow it to cool, carefully remove the stopper *for not more than a second or two to* allow air to enter, and again insert the stopper. (As the flask cools the vapor inside condenses and the pressure drops, which explains why air rushes in when the stopper is removed.)

Dry the flask with a towel to remove the surface water and let it cool to room temperature. Loosen the stopper momentarily to equalize any pressure differences, and reweigh the flask. Read the atmospheric pressure from the barometer.

Repeat the procedure using another 5 mL of your liquid sample.

You may obtain the volume of the flask from your instructor. Alternatively, he may direct you to measure its volume by weighing the flask stoppered and full of water on a top-loading balance. *Do not* fill the flask with water unless specifically told to do so.

When you have completed the experiment, return the flask to the storeroom; do not attempt to wash or clean it in any way.

Data: Molar Mass of a Volatile Liquid

	Trial 1	*Trial 2*
Unknown no.	_____	
Mass of flask and stopper	_____ g	_____ g
Mask of flask, stopper, and condensed vapor	_____ g	_____ g
Mass of flask, stopper, and water (see directions)	_____ g	_____ g
Temperature of boiling water bath	_____ °C	_____ °C
Barometric pressure	_____ mm Hg	_____ mm Hg

Calculations and Results

	Trial 1	*Trial 2*
Pressure of vapor, P	_____ atm	_____ atm
Volume of flask (volume of vapor), V	_____ L	_____ L
Temperature of vapor, T	_____ K	_____ K
Mass of vapor, g	_____ g	_____ g
Number of moles of vapor, n	_____	_____
Molar mass of unknown, as found by substitution into Equation 2	_____ g	_____ g

Advance Study Assignment: Molar Mass of a Volatile Liquid

1. A student weighs an empty flask and stopper and finds the mass to be 55.441 g. She then adds about 5 mL of an unknown liquid and heats the flask in a boiling water bath at 100°C. After all the liquid is vaporized, she removes the flask from the bath, stoppers it, and lets it cool. After it is cool, she momentarily removes the stopper, then replaces it and weighs the flask and condensed vapor, obtaining a mass of 56.039 g. The volume of the flask is known to be 215.8 mL. The barometric pressure in the laboratory that day is 752 mm Hg.

    a. What was the pressure of the vapor in the flask in atm?

    $P =$ _____ atm

    b. What was the temperature of the vapor in K? the volume of the flask in liters?

    $T =$ _____ K $V =$ _____ L

    c. What was the mass of vapor that was present in the flask?

    $g =$ _____ grams

    d. How many moles of vapor are present?

    $n =$ _____ moles

    e. What is the mass of one mole of vapor? (Eq. 2.)

    $MM =$ _____ g/mole

2. How would each of the following procedural errors affect the results to be expected in this experiment? Give your reasoning in each case.

    a. All of the liquid was not vaporized when the flask was removed from the water bath.

    b. The flask was not dried before the final weighing with the condensed vapor inside.

    c. The flask was left open to the atmosphere while it was being cooled, and the stopper was inserted just before the final weighing.

    d. The flask was removed from the bath before the vapor had reached the temperature of the boiling water. All the liquid had vaporized.

Experiment 9

ANALYSIS OF AN ALUMINUM-ZINC ALLOY*

Some of the more active metals will react readily with solutions of strong acids, producing hydrogen gas and a solution of a salt of the metal. Small amounts of hydrogen are commonly prepared by the action of hydrochloric acid on metallic zinc:

$$Zn(s) + 2\,H^+\,(aq) \rightarrow H_2(g) + Zn^{2+}(aq) \tag{1}$$

From this equation it is clear that one mole of zinc produces one mole of hydrogen gas in this reaction. If the hydrogen were collected under known conditions, it would be possible to calculate the mass of zinc in a pure sample by measuring the amount of hydrogen it produced on reaction with acid.

Since aluminum reacts spontaneously with strong acids in a manner similar to that shown by zinc,

$$2\,Al(s) + 6\,H^+(aq) \rightarrow 2\,Al^{3+}(aq) + 3\,H_2(g) \tag{2}$$

we could find the amount of aluminum in a pure sample by measuring the amount of hydrogen produced by its reaction with an acid solution. In this case two moles of aluminum would produce three moles of hydrogen.

Since the amount of hydrogen produced by a gram of zinc is not the same as the amount produced by a gram of aluminum,

$$1 \text{ mole Zn} \rightarrow 1 \text{ mole } H_2, \; 65.4 \text{ g Zn} \rightarrow 1 \text{ mole } H_2, \; 1.00 \text{ g Zn} \rightarrow 0.0153 \text{ mole } H_2 \tag{3}$$

$$2 \text{ moles Al} \rightarrow 3 \text{ moles } H_2, \; 54.0 \text{ g Al} \rightarrow 3 \text{ moles } H_2, \; 1.00 \text{ g Al} \rightarrow 0.0556 \text{ mole } H_2 \tag{4}$$

it is possible to react an alloy of zinc and aluminum of known mass with acid, determine the amount of hydrogen gas evolved, and calculate the percentages of zinc and aluminum in the alloy, using Relations 3 and 4. The object of this experiment is to make such an analysis.

In this experiment you will react a weighed sample of an aluminum-zinc alloy with an excess of acid and collect the hydrogen gas evolved over water (Fig. 9.1). If you measure the volume, temperature, and total pressure of the gas and use the Ideal Gas Law, taking proper account of the pressure of water vapor in the system, you can calculate the number of moles of hydrogen produced by the sample:

$$P_{H_2}V = n_{H_2}RT, \quad n_{H_2} = \frac{P_{H_2}V}{RT} \tag{5}$$

The volume V and the temperature T of the hydrogen are easily obtained from the data. The pressure exerted by the dry hydrogen P_{H_2} requires more attention. The total pressure P of

* W. L. Masterton, J. Chem. Educ. *38*, 558 (1961). For a source of alloy samples, see instructor's manual.

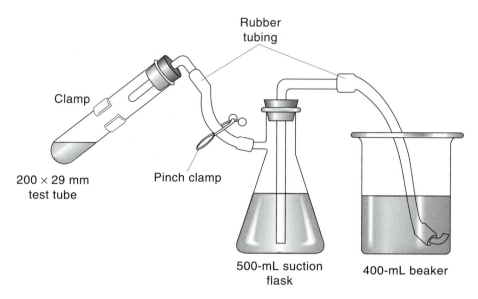

Rubber
tubing

Clamp

200 × 29 mm
test tube

Pinch clamp

500-mL suction
flask

400-mL beaker

FIGURE 9.1

gas in the bottle is, by Dalton's Law, equal to the partial pressure of the hydrogen P_{H_2} plus the partial pressure of the water vapor P_{H_2O}:

$$P = P_{H_2} + P_{H_2O} \tag{6}$$

The water vapor in the bottle is present with liquid water, so the gas is saturated with water vapor; the pressure P_{H_2O} under these conditions is equal to the vapor pressure VP_{H_2O} of water at the temperature of the experiment. This value is constant at a given temperature, and is found in Appendix I at the end of this manual. The total gas pressure P in the flask is very nearly equal to the barometric pressure P_{bar}.*

Substituting these values into (6) and solving for P_{H_2}, we obtain

$$P_{H_2} = P_{bar} - VP_{H_2O} \tag{7}$$

Using (5), you can now calculate n_{H_2}, the number of moles of hydrogen produced by your weighed sample. You can then calculate the percentages of Al and Zn in the sample by properly applying (3) and (4) to your results. For a sample containing g_{Al} grams Al and g_{Zn} grams Zn, it follows that

$$n_{H_2} = (g_{Al} \times 0.0556) + (g_{Zn} \times 0.0153) \tag{8}$$

For a one-gram sample, g_{Al} and g_{Zn} represent the mass fractions of Al and Zn, that is, % Al/100 and % Zn/100. Therefore

$$N_{H_2} = \left(\frac{\% \text{ Al}}{100} \times 0.0556 \right) + \left(\frac{\% \text{ Zn}}{100} \times 0.0153 \right) \tag{9}$$

where N_{H_2} = number of moles of H_2 produced *per gram* of sample.

Since it is also true that

$$\% \text{ Zn} = 100 - \% \text{ Al} \tag{10}$$

---

* In principle a small correction should be made for the difference in heights of the water levels inside and outside the suction flask. In practice the error made by neglecting this effect is smaller than other experimental errors.

Equation (9) can be written in the form

$$N_{H_2} = \left(\frac{\% \text{ Al}}{100} \times 0.0556 \right) + \left(\frac{100 - \% \text{ Al}}{100} \times 0.0153 \right) \qquad (11)$$

We can solve Equation 11 directly for % Al if we know the number of moles of H_2 evolved per gram of sample. To save time in the laboratory and to avoid arithmetic errors, it is highly desirable to prepare in advance a graph giving N_{H_2} as a function of % Al. Then when N_{H_2} has been determined in the experiment, % Al in the sample can be read directly from the graph. Directions for preparing such a graph are given in Problem 1 in the Advance Study Assignment.

Experimental Procedure

Obtain a suction flask, large test tube, stopper assemblies, and a sample of Al-Zn alloy from the stockroom. Assemble the apparatus as shown in Figure 9.1.

Take a gelatin capsule from the supply on the lab bench and weigh it on the analytical balance to ± 0.0001 g. Pour your alloy sample out on a piece of paper and add about half of it to the capsule. If necessary, break up the turnings into smaller pieces by simply tearing them. Cover the capsule and weigh it again. The mass of sample should be between 0.1500 and 0.2500 g. Use care in both weighings, since the sample is small and a small weighing error will produce a large experimental error. Put the remaining alloy back in its container.

Fill the suction flask and beaker about 2/3 full of water. Moisten the stopper on the suction flask and insert it firmly into the flask. Open the pinch clamp and apply suction to the tubing attached to the side arm of the suction flask. Pull water into the flask from the beaker until the water level in the flask is 4 or 5 cm below the side arm. To apply suction, use a suction bulb or a short piece of rubber tubing attached temporarily to the tube that goes through the test tube stopper. Close the pinch clamp to prevent siphoning. The tubing from the beaker to the flask should be full of water, with no air bubbles.

Carefully remove the tubing from the beaker and put the end on the lab bench. As you do this, no water should leak out of the end of the tubing. Pour the water remaining in the beaker into another beaker, letting the 400-mL beaker drain for a second or two. Without drying it, weigh the empty beaker on a top-loading balance to ± 0.1 g. Put the tubing back in this beaker.

Pour 10 mL of 6 M HCl, hydrochloric acid, as measured in your graduated cylinder, into the large test tube. Drop the gelatin capsule into the HCl solution; if it sticks to the tube, poke it down into the acid with your stirring rod. Insert the stopper firmly into the test tube and open the pinch clamp. If a little water goes into the beaker at that point, pour that water out, letting the beaker drain for a second or two.

Within 3 or 4 minutes the acid will eat through the wall of the capsule and begin to react with the alloy. The hydrogen gas that is formed will go into the suction flask and displace water from the flask into the beaker. The volume of water that is displaced will equal the volume of gas that is produced. As the reaction proceeds you will probably observe a dark foam, which contains particles of unreacted alloy. The foam may carry some of the alloy up the tube. Wiggle the tube gently to make sure that all of the alloy gets into the acid solution. The reaction should be over within 5 to 10 minutes. At that time the liquid solution will again be clear, the foam will be essentially gone, the capsule will be all dissolved, and there should be no unreacted alloy. When the reaction is over, close the pinch clamp and take the tubing out of the beaker. Weigh the beaker and the displaced water to ± 0.1 g. Measure the temperature of the water and the barometric pressure.

Pour the acid solution into the waste crock. Reassemble the apparatus and repeat the experiment with the remaining sample of alloy.

Data: Analysis of an Aluminum-Zinc Alloy

	Trial 1	*Trial 2*
Mass of gelatin capsule	_____ g	_____ g
Mass of alloy sample plus capsule	_____ g	_____ g
Mass of empty beaker	_____ g	_____ g
Mass of beaker plus displaced water	_____ g	_____ g
Barometric pressure	_____ mm Hg	
Temperature	_____ °C	

Calculations

	Trial 1	*Trial 2*
Mass of alloy sample	_____ g	_____ g
Mass of displaced water	_____ g	_____ g
Volume of displaced water ($d = 1$ g/mL)	_____ mL	_____ mL
Volume of H_2, V	_____ L	_____ L
Temperature of H_2, T	_____ K	
Vapor pressure of water at T, VP_{H_2O}, from Appendix I	_____ mm Hg	
Pressure of dry H_2, P_{H_2}	_____ mm Hg;	_____ atm
Moles H_2 from sample, n_{H_2}	_____ moles	_____ moles
Moles H_2 per gram of sample, N_{H_2}	_____ moles/g	_____ moles/g
% Al (read from graph)	_____ %	_____ %
Unknown no.	_____	

Advance Study Assignment: Analysis of an Aluminum-Zinc Alloy

1. On the following page, construct a graph of N_{H_2} vs. % Al. To do this, refer to Equation 11 and the discussion preceding it. Note that a plot of N_{H_2} vs. % Al should be a straight line (why?). To fix the position of a straight line it is necessary to locate only two points. The most obvious way to do this is to find N_{H_2} when % Al = 0 and when % Al = 100. If you wish you may calculate some intermediate points (for example, N_{H_2} when % Al = 50, or 20, or 70); all these points should lie on the same straight line.

2. A student obtained the following data in this experiment. Fill in the blanks in the data and make the indicated calculations:

Mass of gelatin capsule	0.1168 g	Temperature, t	21°C
Mass of capsule plus alloy sample	0.2754 g	Temperature, T	_____ K
Mass of alloy sample, m	_____ g	Barometric pressure	746 mm Hg
Mass of empty beaker	141.2 g	Vapor pressure of H_2O at t (Appendix I)	_____ mm Hg
Mass of beaker plus displaced water	307.7 g		
Mass of displaced water	_____ g	Pressure of dry H_2, P_{H_2} (Eq. 7)	_____ mm Hg
Volume of displaced water (density = 1.00 g/mL)	_____ mL	Pressure of dry H_2	_____ atm

Volume, V, of H_2 = Volume of displaced water _____ mL; _____ liters

Find the number of moles of H_2 evolved, n_{H_2} (Eq. 5; V in liters, P_{H_2} in atm, T in K, $R = 0.0821$ L-atm/mole K).

_____ moles H_2

Find N_{H_2}, the number of moles of H_2 per gram of sample (n_{H_2}/m).

_____ moles H_2/g

(continued on following page)

(continued)

Find the % Al in the sample from the graph prepared for Problem 1.

_____ % Al

Find the % Al in the sample by using Equation 11.

_____ % Al

Name ______________________________ Section ______________

ANALYSIS OF AN ALUMINUM-ZINC ALLOY
(Advance Study Assignment)

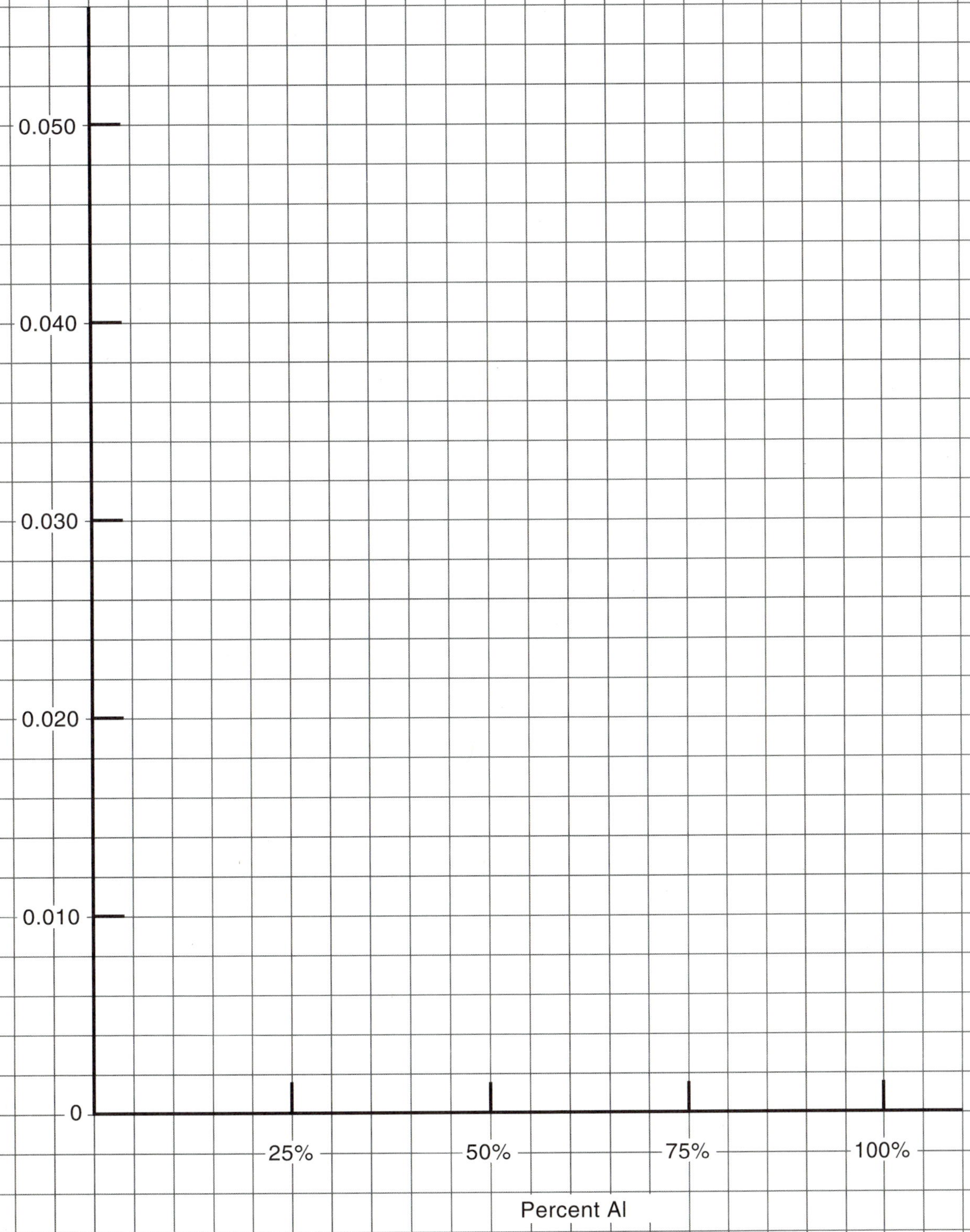

0.050
0.040
0.030
0.020
0.010
0
25%
50%
75%
100%
Percent Al

Experiment 10

THE ATOMIC SPECTRUM OF HYDROGEN

When atoms are excited, either in an electric discharge or with heat, they tend to give off light. The light is emitted only at certain wavelengths that are characteristic of the atoms in the sample. These wavelengths constitute what is called the atomic spectrum of the excited element and reveal much of the detailed information we have regarding the electronic structure of atoms.

Atomic spectra are interpreted in terms of quantum theory. According to this theory, atoms can exist only in certain states, each of which has an associated fixed amount of energy. When an atom changes its state, it must absorb or emit an amount of energy that is just equal to the difference between the energies of the initial and final states. This energy may be absorbed or emitted in the form of light. The emission spectrum of an atom is obtained when excited atoms fall from higher to lower energy levels. Since there are many such levels, the atomic spectra of most elements are very complex.

Light is absorbed or emitted by atoms in the form of photons, each of which has a specific amount of energy, ϵ. This energy is related to the wavelength of light by the equation

$$\epsilon_{photon} = \frac{hc}{\lambda} \tag{1}$$

where h is Planck's constant, 6.62608×10^{-34} joule seconds, c is the speed of light, 2.997925×10^8 meters per second, and λ is the wavelength, in meters. The energy ϵ_{photon} is in joules and is the energy given off by one atom when it jumps from a higher to a lower energy level. Since total energy is conserved, the change in energy of the atom, $\Delta\epsilon_{atom}$, must equal the energy of the photon emitted:

$$\Delta\epsilon_{atom} = \epsilon_{photon} \tag{2}$$

where $\Delta\epsilon_{atom}$ is equal to the energy in the upper level minus the energy in the lower one. Combining Equations 1 and 2, we obtain the relation between the change in energy of the atom and the wavelength of light associated with that change:

$$\Delta\epsilon_{atom} = \epsilon_{upper} - \epsilon_{lower} = \epsilon_{photon} = \frac{hc}{\lambda} \tag{3}$$

The amount of energy in a photon given off when an atom makes a transition from one level to another is very small, of the order of 1×10^{-19} joules. This is not surprising since, after all, atoms are very small particles. To avoid such small numbers, we will work with one mole of atoms, much as we do in dealing with energies involved in chemical reactions. To do this we need only to multiply Equation 3 by Avogadro's number, N:

Let

$$N\Delta\epsilon = \Delta E = N\epsilon_{upper} - N\epsilon_{lower} = E_{upper} - E_{lower} = \frac{Nhc}{\lambda}$$

Substituting the values for N, h, and c, and expressing the wavelength in nanometers rather than meters (1 meter = 1×10^9 nanometers), we obtain an equation relating energy change in kilojoules per mole of atoms to the wavelength of photons associated with such a change:

$$\Delta E = \frac{6.02214 \times 10^{23} \times 6.62608 \times 10^{-34} \text{ J sec} \times 2.997925 \times 10^8 \text{ m/sec}}{\lambda \text{ (in nm)}}$$

$$\times \frac{1 \times 10^9 \text{ nm}}{1 \text{ m}} \times \frac{1 \text{ kJ}}{1000 \text{ J}}$$

$$\Delta E = E_{upper} - E_{lower} = \frac{1.19627 \times 10^5 \text{ kJ/mole}}{\lambda \text{ (in nm)}} \quad \text{or} \quad \lambda \text{ (in nm)} = \frac{1.19627 \times 10^5}{\Delta E \text{ (in kJ/mole)}} \quad (4)$$

Equation 4 is useful in the interpretation of atomic spectra. Say, for example, we study the atomic spectrum of sodium and find that the wavelength of the strong yellow line is 589.16 nm (see Fig. 10.1). This line is known to result from a transition between two of the three lowest levels in the atom. The energies of these levels are shown in the figure. To make the determination of the levels which give rise to the 589.16 nm line, we note that there are three possible transitions, shown by downward arrows in the figure. We find the wavelengths associated with those transitions by first calculating ΔE ($E_{upper} - E_{lower}$) for each transition. Knowing ΔE we calculate λ by Equation 4. Clearly, the II $\rightarrow$ I transition is the source of the yellow line in the spectrum.

The simplest of all atomic spectra is that of the hydrogen atom. In 1886 Balmer showed that the lines in the spectrum of the hydrogen atom had wavelengths that could be expressed by a rather simple equation. Bohr, in 1913, explained the spectrum on a theoretical basis with his famous model of the hydrogen atom. According to Bohr's theory, the energies allowed to a hydrogen atom are given by the equation

$$\epsilon_n = \frac{-B}{n^2} \quad (5)$$

where B is a constant predicted by the theory and n is an integer, 1, 2, 3, . . . , called a quantum number. It has been found that all the lines in the atomic spectrum of hydrogen can be associated with energy levels in the atom which are predicted with great accuracy by

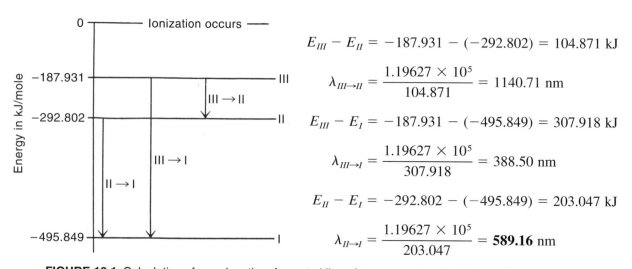

$$E_{III} - E_{II} = -187.931 - (-292.802) = 104.871 \text{ kJ}$$

$$\lambda_{III \to II} = \frac{1.19627 \times 10^5}{104.871} = 1140.71 \text{ nm}$$

$$E_{III} - E_{I} = -187.931 - (-495.849) = 307.918 \text{ kJ}$$

$$\lambda_{III \to I} = \frac{1.19627 \times 10^5}{307.918} = 388.50 \text{ nm}$$

$$E_{II} - E_{I} = -292.802 - (-495.849) = 203.047 \text{ kJ}$$

$$\lambda_{II \to I} = \frac{1.19627 \times 10^5}{203.047} = \mathbf{589.16} \text{ nm}$$

FIGURE 10.1 Calculation of wavelengths of spectral lines from energy levels of the sodium atom.

Bohr's equation. When we write Equation 5 in terms of a mole of H atoms, and substitute the numerical value for B, we obtain

$$E_n = \frac{-1312.04}{n^2} \text{ kilojoules per mole, } n = 1, 2, 3, \ldots \qquad (6)$$

Using Equation 6 you can calculate, very accurately indeed, the energy levels for hydrogen. Transitions between these levels give rise to the wavelengths in the atomic spectrum of hydrogen. These wavelengths are also known very accurately. Given both the energy levels and the wavelengths, it is possible to determine the actual levels associated with each wavelength. In this experiment your task will be to make determinations of this type for the observed wavelengths in the hydrogen atomic spectrum that are listed in Table 10.1.

Table 10.1   Some Wavelengths (in nm) in the Spectrum of the
Hydrogen Atom as Measured in a Vacuum

Wavelength	Assignment $n_{hi} \rightarrow n_{lo}$	Wavelength	Assignment $n_{hi} \rightarrow n_{lo}$	Wavelength	Assignment $n_{hi} \rightarrow n_{lo}$
97.25	_____	410.29	_____	1005.2	_____
102.57	_____	434.17	_____	1094.1	_____
121.57	_____	486.27	_____	1282.2	_____
389.02	_____	656.47	_____	1875.6	_____
397.12	_____	954.86	_____	4052.3	_____

Experimental Procedure

There are several ways we might analyze an atomic spectrum, given the energy levels of the atom involved. A simple and effective method is to calculate the wavelengths of some of the lines arising from transitions between some of the lower energy levels, and see if they match those that are observed. We shall use this method in our experiment. All the data are good to at least five significant figures, so by using electronic calculators you should be able to make very accurate determinations.

A. Calculations of the Energy Levels of the Hydrogen Atom

Given the expression for E_n in Equation 6, it is possible to calculate the energy for each of the allowed levels of the H atom starting with $n = 1$. Using your calculator, calculate the energy in kJ/mole of each of the 10 lowest levels of the H atom. Note that the energies are all negative, so that the *lowest* energy will have the *largest* allowed negative value. Enter these values in the table of energy levels, Table 10.2. On the energy level diagram provided, plot along the y axis each of the six lowest energies, drawing a horizontal line at the allowed level and writing the value of the energy alongside the line near the y axis. Write the quantum number associated with the level to the right of the line.

* On this experiment you may work without safety glasses unless otherwise directed by your instructor.

B. Calculation of the Wavelengths of the Lines in the Hydrogen Spectrum

The lines in the hydrogen spectrum all arise from jumps made by the atom from one energy level to another. The wavelengths in nm of these lines can be calculated by Equation 4, where ΔE is the difference in energy in kJ/mole between any two allowed levels. For example, to find the wavelength of the spectral line associated with a transition from the $n = 2$ level to the $n = 1$ level, calculate the difference, ΔE, between the energies of those two levels. Then substitute ΔE into Equation 4 to obtain this wavelength in nanometers.

Using the procedure we have outlined, calculate the wavelengths in nm of all the lines we have indicated in Table 10.3. That is, calculate the wavelengths of all the lines that can arise from transitions between any two of the six lowest levels of the H atom. Enter these values in Table 10.3.

C. Assignment of Observed Lines in the Hydrogen Spectrum

Compare the wavelengths you have calculated with those listed in Table 10.1. If you have made your calculations properly, your wavelengths should match, within the error of your calculation, several of those that are observed. On the line opposite each wavelength in Table 10.1, write the quantum numbers of the upper and lower states for each line whose origin you can recognize by comparison of your calculated values with the observed values. On the energy level diagram, draw a vertical arrow pointing down (light is emitted, $\Delta E < 0$) between those pairs of levels that you associate with any of the observed wavelengths. By each arrow write the wavelength of the line originating from that transition.

There are a few wavelengths in Table 10.1 that have not yet been calculated. Enter those wavelengths in Table 10.4. By assignments already made and by an examination of the transitions you have marked on the diagram, deduce the quantum states that are likely to be associated with the as yet unassigned lines. This is perhaps most easily done by first calculating the value of ΔE, which is associated with a given wavelength. Then find two values of E_n whose difference is equal to ΔE. The quantum numbers for the two E_n states whose energy difference is ΔE will be the ones that are to be assigned to the given wavelength. When you have found n_{hi} and n_{lo} for a wavelength, write them in Table 10.1 and Table 10.4; continue until all the lines in the table have been assigned.

D. The Balmer Series

This is the most famous series in the atomic spectrum of hydrogen. The lines in this series are the only ones in the spectrum that occur in the visible region. Your instructor may have a hydrogen source tube and a spectroscope with which you may be able to observe some of the lines in the Balmer series. In the Data and Calculations section are some questions you should answer relating to this series.

Data and Calculations: The Atomic Spectrum of Hydrogen

A. The Energy Levels of the Hydrogen Atom

Energies are to be calculated from Equation 6 for the 10 lowest energy states.

Table 10.2

Quantum Number, n	*Energy,* E_n, *in kJ/mole*	*Quantum Number,* n	*Energy,* E_n, *in kJ/mole*
_____	_____	_____	_____
_____	_____	_____	_____
_____	_____	_____	_____
_____	_____	_____	_____
_____	_____	_____	_____

B. Calculation of Wavelengths in the Spectrum of the H Atom

In the upper half of each box write ΔE, the difference in energy in kJ/mole between $E_{n_{hi}}$ and $E_{n_{lo}}$. In the lower half of the box, write λ in nm associated with that value of ΔE.

Table 10.3

n_{higher}	6	5	4	3	2	1
n_{lower}						
1						
2						
3						
4						
5						

$$\Delta E = E_{n_{hi}} - E_{n_{lo}}$$

$$\lambda \text{ (nm)} = \frac{1.19627 \times 10^5}{\Delta E}$$

(continued on following page)

(continued)

C. Assignment of Wavelengths

1. As directed in the procedure, assign n_{hi} and n_{lo} for each wavelength in Table 10.1 which corresponds to a wavelength calculated in Table 10.3.

2. List below any wavelengths you cannot yet assign.

Table 10.4

Wavelength λ Observed	ΔE Transition	Probable Transition $n_{ni} \rightarrow n_{lo}$	λ Calculated in nm (Eq. 4)
————	————	————	————
————	————	————	————
————	————	————	————
————	————	————	————

D. The Balmer Series

1. When Balmer found his famous series for hydrogen in 1886, he was limited experimentally to wavelengths in the visible and near ultraviolet regions from 250 nm to 700 nm, so all the lines in his series lie in that region. On the basis of the entries in Table 10.3 and the transitions on your energy level diagram, what common characteristic do the lines in the Balmer Series have?

What would be the longest possible wavelength for a line in the Balmer series?

$\lambda =$ _____ nm

What would be the shortest possible wavelength that a line in the Balmer series could have? Hint: What is the largest possible value of ΔE to be associated with a line in the Balmer series?

$\lambda =$ _____ nm

Fundamentally, why would any line in the hydrogen spectrum between 250 nm and 700 nm belong to the Balmer series? Hint: On the energy level diagram note the range of possible values of ΔE for transitions to the $n = 1$ level and to the $n = 3$ level. Could a spectral line involving a transition to the $n = 1$ level have a wavelength in the range indicated?

(continued on following page)

The Ionization Energy of Hydrogen

2. In the normal hydrogen atom the electron is in its lowest energy state, which is called the ground state of the atom. The maximum electronic energy that a hydrogen atom can have is 0 kJ/mole, at which point the electron would essentially be removed from the atom and it would become a H^+ ion. How much energy in kilojoules per mole does it take to ionize an H atom?

_____ kJ/mole

The ionization energy of hydrogen is often expressed in units other than kJ/mole. What would it be in joules per atom? in electron volts per atom? (1 ev $= 1.602 \times 10^{-19}$ J)

_____ J/atom; _____ ev/atom

(The energy level diagram to be completed in Part A is on the following page.)

The Atomic Spectrum of Hydrogen
Energy Level Diagram

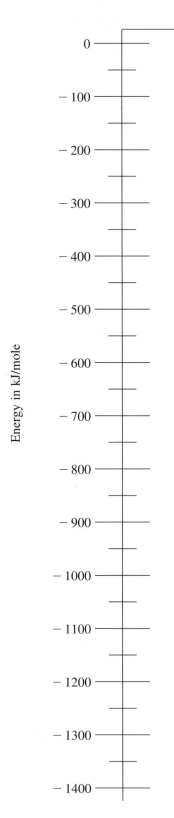

Energy in kJ/mole

0

− 100

− 200

− 300

− 400

− 500

− 600

− 700

− 800

− 900

− 1000

− 1100

− 1200

− 1300

− 1400

Advance Study Assignment: The Atomic Spectrum of Hydrogen

1. The helium ion, He$^+$, has energy levels similar to those of the hydrogen atom, since both species have only one electron. The energy levels of the He$^+$ ion are given by the equation

$$E_n = -\frac{5248.16}{n^2} \text{ kJ/mole} \quad n = 1, 2, 3, \ldots$$

   a. Calculate the energies in kJ/mole for the four lowest energy levels of the He$^+$ ion.

$$E_1 = \underline{\hspace{2cm}} \text{ kJ/mole}$$

$$E_2 = \underline{\hspace{2cm}} \text{ kJ/mole}$$

$$E_3 = \underline{\hspace{2cm}} \text{ kJ/mole}$$

$$E_4 = \underline{\hspace{2cm}} \text{ kJ/mole}$$

   b. One of the most important transitions for the He$^+$ ion involves a jump from the $n = 2$ to the $n = 1$ level. ΔE for this transition equals $E_2 - E_1$, where these two energies are obtained as in Part a. Find the value of ΔE in kJ/mole. Find the wavelength in nm of the line emitted when this transition occurs; use Equation 4 to make the calculation.

$$\Delta E = \underline{\hspace{2cm}} \text{ kJ/mole}; \lambda = \underline{\hspace{2cm}} \text{ nm}$$

   c. Three of the strongest lines in the He$^+$ ion spectrum are observed at the following wavelengths: (1) 121.57 nm; (2) 164.12 nm; (3) 468.90 nm. Find the quantum numbers of the initial and final states for the transitions that give rise to these three lines. Do this by calculating, using Equation 4, the wavelengths of lines that can originate from transitions involving any two of the four lowest levels. You calculated one such wavelength in Part b. Make similar calculations with the other possible pairs of levels. When a calculated wavelength matches an observed one, write down n_{hi} and n_{lo} for that line. Continue until you have assigned all three of the lines. Make your calculations on the other side of this page.

   (1) _____ $\rightarrow$ ____   (2) _____ $\rightarrow$ ____   (3) _____ $\rightarrow$ ____

THE ALKALINE EARTHS AND THE HALOGENS—TWO FAMILIES IN THE PERIODIC TABLE

The Periodic Table arranges the elements in order of increasing atomic number in horizontal rows of such length that elements with similar properties recur periodically; that is, they fall directly beneath each other in the Table. The elements in a given vertical column are referred to as a family or group. The physical and chemical properties of the elements in a given family change gradually as one goes from one element in the column to the next. By observing the trends in properties the elements can be arranged in the order in which they appear in the Periodic Table. In this experiment we will study the properties of the elements in two families in the Periodic Table, the alkaline earths (Group 2) and the halogens (Group 7).

The alkaline earths are all moderately reactive metals and include barium, beryllium, calcium, magnesium, radium, and strontium. (Since beryllium compounds are rarely encountered and often very poisonous, and radium compounds are highly radioactive, we will not include these two elements in this experiment.) All the alkaline earths exist in their compounds and in solution as M^{2+} cations (Mg^{2+}, Ca^{2+}, etc.). If a solution containing one of these cations is mixed with one containing an anion (CO_3^{2-}, SO_4^{2-}, IO_3^-, etc.), an alkaline earth salt will precipitate if the compound containing those two ions is insoluble.

For example:

$$M^{2+}(aq) + SO_4^{2-}(aq) \rightarrow MSO_4(s) \qquad \text{if } MSO_4 \text{ is insoluble} \qquad (1a)$$

$$M^{2+}(aq) + 2 IO_3^-(aq) \rightarrow M(IO_3)_2(s) \quad \text{if } M(IO_3)_2 \text{ is insoluble} \qquad (1b)$$

We would expect, and indeed observe, that the solubilities of the salt of the alkaline earth cations with any one of the given anions show a smooth trend consistent with the order of the cations in the Periodic Table. That is, as we go from one end of the alkaline earth family to the other, the solubilities of, say, the sulfate salts either gradually increase or decrease. Similar trends exist for the carbonates, oxalates, and iodates formed by those cations. By determining such trends in this experiment, you will be able to confirm the order of the alkaline earths in the Periodic Table.

The elementary halogens are also relatively reactive. They include astatine, bromine, chlorine, fluorine, and iodine. We will not study astatine and fluorine in this experiment, since the former is radioactive and the latter is too reactive to be safe. Unlike the alkaline earths, the halogen atoms tend to gain electrons, forming X^- anions (Cl^-, Br^-, etc.). Because of this property, the halogens are oxidizing agents, species that tend to oxidize (remove electrons from) other species. An interesting and simple example of the sort of reaction that may occur arises when a solution containing a halogen (Cl_2, Br_2, I_2) is mixed with a solution containing a halide ion (Cl^-, Br^-, I^-). Taking X_2 to be the halogen, and Y^- to be a halide ion, the following reaction may occur, in which another halogen, Y_2, is formed:

$$X_2(aq) + 2 Y^-(aq) \rightarrow 2 X^-(aq) + Y_2(aq) \qquad (2)$$

The reaction will occur if X_2 is a better oxidizing agent than Y_2, since then X_2 can produce Y_2 by removing electrons from the Y^- ions. If Y_2 is a better oxidizing agent than X_2, Reaction 2 will not proceed but will be spontaneous in the opposite direction.

In this experiment we will mix solutions of halogens and halide ions to determine the relative oxidizing strengths of the halogens. These strengths show a smooth variation as one goes from one halogen to the next in the Periodic Table. We will be able to tell if a reaction occurs by the colors we observe. In water, and particularly in some organic solvents, the halogens have characteristic colors. The halide ions are colorless in water solution and insoluble in organic solvents. Bromine (Br_2) in hexane, C_6H_{14}(HEX), is orange, while Cl_2 and I_2 in that solvent have quite different colors.

Say, for example, we shake a water solution of Br_2 with a little hexane, which is lighter than and insoluble in water. The Br_2 is much more soluble in HEX than in water and goes into the HEX layer, giving it an orange color. To that mixture we add a solution containing a halide ion, say Cl^- ion, and mix well. If Br_2 is a better oxidizing agent than Cl_2, it will take electrons from the chloride ions and will be converted to bromide, Br^-, ions; the reaction would be

$$Br_2(aq) + 2\ Cl^-(aq) \rightarrow 2\ Br^-(aq) + Cl_2(aq) \tag{3}$$

If the reaction occurs, the color of the HEX layer will of necessity change, since Br_2 will be used up and Cl_2 will form. The color of the HEX layer will go from orange to that of a solution of Cl_2 in HEX. If the reaction does *not* occur, the color of the HEX layer will remain orange. By using this line of reasoning, and by working with the possible mixtures of halogens and halide ions, you should be able to arrange the halogens in order of increasing oxidizing power, which must correspond to their order in the Periodic Table.

One difficulty that you may have in this experiment involves terminology rather than actual chemistry. You must learn to distinguish the halogen *elements* from the halide *ions,* since the two kinds of species are not at all the same, even though their names are similar:

Elementary Halogens	*Halide Ions*
Bromine, Br_2	Bromide ion, Br^-
Chlorine, Cl_2	Chloride ion, Cl^-
Iodine, I_2	Iodide ion, I^-

The *halogens* are molecular substances and oxidizing agents, and all have odors. They are only slightly soluble in water and are much more soluble in HEX, where they have distinct colors. The *halide ions* exist in solution only in water, have no color or odor, and are *not* oxidizing agents. They do not dissolve in HEX.

Given the solubility properties of the alkaline earth cations, and the oxidizing power of the halogens, it is possible to develop a systematic procedure for determining the presence of any Group 2 cation and any Group 7 anion in a solution. In the last part of this experiment you will be asked to set up such a procedure and use it to establish the identity of an unknown solution containing a single alkaline earth halide.

Experimental Procedure

I. Relative Solubilities of Some Salts of the Alkaline Earths

To each of four small test tubes add about 1 mL (approximately 12 drops) of 1 M H_2SO_4. Then add 1 mL of 0.1 M solutions of the nitrate salts of barium, calcium, magnesium, and strontium to those tubes, one solution to a tube. Stir each mixture with your glass stirring rod, rinsing the rod in a beaker of distilled water between stirs. Record your results on the solubilities of the sulfates of the alkaline earths in the Table, noting whether a precipitate forms, and any characteristics (such as color, amount, size of particles, and settling tendencies) that might distinguish it.

Rinse out the test tubes, and to each add 1 mL 1 M Na_2CO_3. Then add 1 mL of the solutions of the alkaline earth salts, one solution to a tube, as before. Record your observations on the solubility properties of the carbonates of the alkaline earth cations. Rinse out the tubes, and test for the solubilities of the oxalates of these cations, using 0.25 M $(NH_4)_2C_2O_4$ as the precipitating reagent. Finally, determine the relative solubilities of the iodates of the alkaline earths, using 1 mL 0.1 M KIO_3 as the test reagent.

II. Relative Oxidizing Powers of the Halogens

In a small test tube place a few milliliters of bromine-saturated water and add 1 mL of hexane. Stopper the test tube and shake until the bromine color is mostly in the HEX layer. (**C A U T I O N S:** *Avoid breathing the halogen vapors. Don't use your finger to stopper the tube, since a halogen solution can give you a bad chemical burn.*) Repeat the experiment using chlorine water and iodine water with separate samples of HEX, noting any color changes as the bromine, chlorine, and iodine are extracted from the water layer into the HEX layer.

To each of three small test tubes add 1 mL bromine water and 1 mL HEX. Then add 1 mL 0.1 M NaCl to the first test tube, 1 mL 0.1 M NaBr to the second, and 1 mL 0.1 M NaI to the third. Stopper each tube and shake it. Note the color of the HEX phase above each solution. If the color is not that of Br_2 in HEX, a reaction indeed occurred, and Br_2 oxidized that anion, producing the halogen. In such a case, Br_2 is a stronger oxidizing agent than the halogen that was produced.

Rinse out the tubes, and this time add 1 mL chlorine water and 1 mL HEX to each tube. Then add 1 mL of the 0.1 M solutions of the sodium halide salts, one solution to a tube, as before. Stopper each tube and shake, noting the color of the HEX layer after shaking. Depending on whether the color is that of Cl_2 in HEX or not, decide whether Cl_2 is a better oxidizing agent than Br_2 or I_2. Again, rinse out the tubes, and add 1 mL iodine water and 1 mL HEX to each. Test each tube with 1 mL of a sodium halide salt solution, and determine whether I_2 is able to oxidize Cl^- or Br^- ions. Record all your observations in the Table.

III. Identification of an Alkaline Earth Halide

Your observations on the solubility properties of the alkaline earth cations should allow you to develop a method for determining which of those cations is present in a solution containing one Group 2 cation and no other cations. The method will involve testing samples of the solution with one or more of the reagents you used in Part I. Indicate on the Data page how you would proceed.

In a similar way you can determine which halide ion is present in a solution containing only one such anion and no others. There you will need to test a solution of an oxidizing halogen with your unknown to see how the halide ion is affected. From the behavior of the halogen-halide ion mixtures you studied in Part II you should be able to identify easily the particular halide that is present. Describe your method on the Data page, obtain an unknown solution of an alkaline earth halide, and then use your procedure to determine the cation and anion that it contains.

IV. Microscale Procedure for Determining Solubilities of Alkaline Earth Salts (Optional)

Your instructor may have you carry out Part I of this experiment by a microscale approach. This method uses much smaller amounts of reagents. Plastic well plates are employed as containers, and reagents are measured out with small Beral pipettes.

Using Beral pipettes, add four drops 0.1 M $Ba(NO_3)_2$, barium nitrate, to wells A1-A4, four drops to each well. Similarly, add four drops 0.1 M $Ca(NO_3)_2$, calcium nitrate, to wells B1-B4; four drops of 0.1 M $Mg(NO_3)_2$, magnesium nitrate, to wells C1-C4, and four drops 0.1 M $Sr(NO_3)_2$, strontium nitrate, to wells D1-D4.

Then, with another Beral pipette, add four drops 1 M H_2SO_4, sulfuric acid, to wells A1-D1. In the Table, record your results on the solubilities of the sulfates of the alkaline earths. Note whether a precipitate formed, and any characteristics, such as amount, size of particles, and cloudiness, which might distinguish it.

With a different Beral pipette, add four drops of 1 M Na_2CO_3, sodium carbonate, to wells A2-D2. Record your observations on the solubilities of the carbonates of the alkaline earths. Then carry out the same sort of tests with 0.25 M $(NH_4)_2C_2O_4$, ammonium oxalate, in wells A3-D3, and finally with 0.1 M KIO_3, potassium iodate, in wells A4-D4. Note all of your observations in the Table.

SAMPLE DISPOSAL. Dispose of the reaction products from this experiment as directed by your instructor.

Data and Observations: The Alkaline Earths and the Halogens

I. Solubilities of Salts of the Alkaline Earths

	1 M H_2SO_4	1 M Na_2CO_3	0.25 M $(NH_4)_2C_2O_4$	0.1 M KIO_3
$Ba(NO_3)_2$				
$Ca(NO_3)_2$				
$Mg(NO_3)_2$				
$Sr(NO_3)_2$				

Key: P = precipitate forms; S = no precipitate

Note any distinguishing characteristics of precipitate, such as amount and degree of cloudiness.

Consider the relative solubilities of the Group 2 cations in the various precipitating reagents. On the basis of the trends you observed, list the four alkaline earths in the order in which they should appear in the Periodic Table. *Start with the one which forms the most soluble oxalate.*

most soluble _____ _____ _____ _____ least soluble

Why did you arrange the elements as you did? Is the order consistent with the properties of the cations in all of the participating reagents?

II. Relative Oxidizing Powers of the Halogens

a. Color of the halogen in solution:

	Br₂	*Cl₂*	*I₂*
Water	_____	_____	_____
HEX	_____	_____	_____

(continued on following page)

(continued)

b. Reactions between halogens and halides:

	Br$^-$	Cl$^-$	I$^-$
Br$_2$			
Cl$_2$			
I$_2$			

State initial and final colors of HEX layer. R = reaction occurs; NR = no reaction occurs.

Rank the halogens in order of their increasing oxidizing power.

weakest _____   _____   _____ strongest

Is this their order in the Periodic Table?

III. Identification of an Alkaline Earth Halide

Procedure for identifying the Group 2 cation:

Procedure for identifying the Group 7 anion:

Observations on unknown alkaline earth halide solution:

Cation present _____

Anion present _____

Unknown no. _____

Advance Study Assignment: The Alkaline Earths and the Halogens

1. All of the common noble gases are monatomic and low-boiling. Their boiling points in °C are: Ne, − 245; Ar, − 186; Kr, − 152; Xe, − 107. Using the Periodic Table, predict as best you can the molecular formula and boiling point of radon, Rn, the only radioactive element in this family.

————————   ———————— °C

2. Substances *A, B,* and *C* can all act as oxidizing agents. In solution, *A* is green, *B* is yellow, and *C* is red. In the reactions in which they participate, they are reduced to A^-, B^-, and C^- ions, all of which are colorless. When a solution of *A* is mixed with one containing C^- ions, the color changes from green to red.

Which species is oxidized? —————————

Which is reduced? —————————

When a solution of *A* is mixed with one containing B^- ions, the color remains green.

Is *A* a better oxidizing agent than *C*? —————————

Is *A* a better oxidizing agent than *B*? —————————

Arrange *A, B,* and *C* in order of increasing strengths as oxidizing agents.

3. You are given an unknown, colorless, solution that may contain only one salt from the following set: NaA, NaB, NaC. In solution each salt dissociates completely into the Na^+ ion and the anion A^-, B^-, or C^-, whose properties are given in Problem 2. The Na^+ ion is effectively inert. Given the availability of solutions of *A, B,* and *C*, develop a simple procedure for identifying the salt that is present in your unknown. Use the other side of this page.

THE GEOMETRICAL STRUCTURE OF MOLECULES—AN EXPERIMENT USING MOLECULAR MODELS

Many years ago it was observed that in many of its compounds the carbon atom formed four chemical linkages to other atoms. As early as 1870, graphic formulas of carbon compounds were drawn as shown:

$$
\begin{array}{ccc}
& \mathrm{H} & \\
& | & \\
\mathrm{H}-&\mathrm{C}&-\mathrm{H} \\
& | & \\
& \mathrm{H} &
\end{array}
\qquad
\begin{array}{cc}
\mathrm{H} & \mathrm{H} \\
| & | \\
\mathrm{C} &= \mathrm{C} \\
| & | \\
\mathrm{H} & \mathrm{H}
\end{array}
$$

$$\text{methane} \qquad \text{ethylene}$$

Although such drawings as these would imply that the atom-atom linkages, indicated by valence strokes, lie in a plane, chemical evidence, particularly the existence of only one substance with the graphic formula

$$
\begin{array}{ccc}
& \mathrm{Cl} & \\
& | & \\
\mathrm{H}-&\mathrm{C}&-\mathrm{Cl} \\
& | & \\
& \mathrm{H} &
\end{array}
$$

requires that the linkages be directed toward the corners of a tetrahedron, at the center of which is the carbon atom.

The physical significance of the chemical linkages between atoms, expressed by the lines or valence strokes in molecular structure diagrams, became evident soon after the discovery of the electron. In 1916 in a classic paper, G. N. Lewis suggested, on the basis of chemical evidence, that the single bonds in graphic formulas involve two electrons and that an atom tends to hold eight electrons in its outermost or valence shell.

Lewis' proposal that atoms generally have eight electrons in their outer shells proved to be extremely useful and has come to be known as the octet rule. It can be applied to many atoms, but is particularly important in the treatment of covalent compounds of atoms in the second row of the Periodic Table. For atoms such as carbon, oxygen, nitrogen, and fluorine, the eight valence electrons occur in pairs that occupy tetrahedral positions around the central atom core. Some of the electron pairs do not participate directly in chemical bonding and are called unshared or nonbonding pairs; however, the structures of compounds containing such unshared pairs reflect the tetrahedral arrangement of the four pairs of valence shell electrons. In the H_2O molecule, which obeys the octet rule, the four pairs of electrons around the central oxygen atom occupy essentially tetrahedral positions; there are two unshared nonbonding pairs and two bonding pairs that are shared by the O atom and the two H atoms. The H—O—H bond angle is nearly but not exactly tetrahedral since the properties of shared and unshared pairs of electrons are not exactly alike.

$$\text{H} \overset{\cdot\cdot\text{O}\cdot\cdot}{\diagup \diagdown} \text{H}$$

Most molecules obey the octet rule. Essentially, all organic molecules obey the rule, and so do most inorganic molecules and ions. For species that obey the octet rule it is possible to draw electron-dot, or Lewis, structures. The previous drawing of the H_2O molecule is an example of a Lewis structure. Here are several others:

$$\begin{array}{ccc}
:\ddot{\text{C}}\text{l}: \\
| \\
\text{H}-\text{C}-\text{H} \\
| \\
:\ddot{\text{C}}\text{l}:
\end{array}
\qquad
\begin{array}{c}
\text{H}-\ddot{\text{N}}-\text{H} \\
| \\
\text{H}
\end{array}
\qquad
:\ddot{\text{O}}-\text{H}^-
\qquad
\begin{array}{ccc}
\text{H} & & \text{H} \\
| & & | \\
\text{H}-\text{C}-\text{C}-\text{H} \\
| & & | \\
\text{H} & & \text{H}
\end{array}
\qquad
\begin{array}{ccc}
\text{H} & & \text{H} \\
| & & | \\
\text{H}-\text{C}=\text{C}-\text{H} \\
& & \\
\text{H} & & \text{H}
\end{array}$$

In each of the above structures there are eight electrons around each atom (except for H atoms, which always have two electrons). There are two electrons in each bond. When counting electrons in these structures, one considers the electrons in a bond between two atoms as belonging to the atom under consideration. In the CH_2Cl_2 molecule just above, for example, the Cl atoms each have eight electrons, including the two in the single bond to the C atom. The C atom also has eight electrons, two from each of the four bonds to that atom. The bonding and nonbonding electrons in Lewis structures are all from the *outermost* shells of the atoms involved, and are the so-called valence electrons of those atoms. For the main group elements, the number of valence electrons in an atom is equal to the group number of the element in the Periodic Table. Carbon, in Group 4, has four valence electrons in its atoms; hydrogen, in Group 1, has one; chlorine, in Group 7, has seven valence electrons. In an octet rule structure the valence electrons from all the atoms are arranged in such a way that each atom, except hydrogen, has eight electrons.

Often it is quite easy to construct an octet rule structure for a molecule. Given that an oxygen atom has six valence electrons (Group 6) and a hydrogen atom has one, it is clear that one O and two H atoms have a total of eight valence electrons; the octet rule structure for H_2O, which we discussed earlier, follows by inspection. Structures like that of H_2O, involving only single bonds and nonbonding electron pairs, are common. Sometimes, however, there is a "shortage" of electrons; that is, it is not possible to construct an octet rule structure in which all the electron pairs are either in single bonds or are nonbonding. C_2H_4 is a typical example of such a species. In such cases, octet rule structures can often be made in which two atoms are bonded by two pairs, rather than one pair, of electrons. The two pairs of electrons form a double bond. In the C_2H_4 molecule, shown above, the C atoms each get four of their electrons from the double bond. The assumption that electrons behave this way is supported by the fact that the C=C double bond is both shorter and stronger than the C—C single bond in the C_2H_6 molecule (see above). Double bonds, and triple bonds, occur in many molecules, usually between C, O, N, and/or S atoms.

Lewis structures can be used to predict molecular and ionic geometries. All that is needed is to assume that the four pairs of electrons around each atom are arranged tetrahedrally. We have seen how that assumption leads to the correct geometry for H_2O. Applying the same principle to the species whose Lewis structures we listed earlier, we would predict, correctly, that the CH_2Cl_2 molecule would be tetrahedral (roughly anyway), that NH_3 would be pyramidal (with the nonbonding electron pair sticking up from the pyramid made from the atoms), that the bond angles in C_2H_6 are all tetrahedral, and that the C_2H_4 molecule is planar (the two bonding pairs in the double bond are in a sort of banana bonding arrangement above and below the plane of the molecule). In describing molecular geometry we indicate the positions of the atomic nuclei, not the electrons. The NH_3 molecule is pyramidal, not tetrahedral.

It is also possible to predict polarity from Lewis structures. Polar molecules have their center of positive charge at a different point than their center of negative charge. This separation of charges produces a dipole moment in the molecule. Covalent bonds between different kinds of atoms are polar; all heteronuclear diatomic molecules are polar. In some molecules the polarity from one bond may be canceled by that from others. Carbon dioxide, CO_2, which is linear, is a nonpolar molecule. Methane, CH_4, which is tetrahedral, is also nonpolar. Among the species whose Lewis structures we have listed, we find that H_2O, CH_2Cl_2, NH_3, and OH^- are polar. C_2H_6 and C_2H_4 are nonpolar.

For some molecules with a given molecular formula, it is possible to satisfy the octet rule with different atomic arrangements. A simple example would be

$$
\begin{array}{ccc}
\text{H}\quad\text{H} & & \text{H}\qquad\text{H}\\
|\quad\;| & & |\qquad\;|\\
\text{H}-\text{C}-\text{C}-\ddot{\text{O}}-\text{H} \quad\text{and}\quad & & \text{H}-\text{C}-\ddot{\text{O}}-\text{C}-\text{H}\\
|\quad\;| & & |\qquad\;|\\
\text{H}\quad\text{H} & & \text{H}\qquad\text{H}
\end{array}
$$

The two molecules are called isomers of each other, and the phenomenon is called isomerism. Although the molecular formulas of both substances are the same, C_2H_6O, their properties differ markedly because of their different atomic arrangements.

Isomerism is very common, particularly in organic chemistry, and when double bonds are present, isomerism can occur in very small molecules:

$$
\begin{array}{ccccc}
\text{H}\diagdown\;\;\diagup\text{H} & & \ddot{\text{Cl}}\diagdown\;\;\diagup\text{H} & & \ddot{\text{Cl}}\diagdown\;\;\diagup\text{H}\\
\text{C}=\text{C} & \text{and} & \text{C}=\text{C} & \text{and} & \text{C}=\text{C}\\
\ddot{\text{Cl}}\diagup\;\;\diagdown\ddot{\text{Cl}} & & \text{H}\diagup\;\;\diagdown\ddot{\text{Cl}} & & \ddot{\text{Cl}}\diagup\;\;\diagdown\text{H}
\end{array}
$$

The first two isomers result from the fact that there is no rotation around a double bond, although such rotation can occur around single bonds. The third isomeric structure cannot be converted to either of the first two without breaking bonds.

With certain molecules, given a fixed atomic geometry, it is possible to satisfy the octet rule with more than one bonding arrangement. The classic example is benzene, whose molecular formula is C_6H_6:

These two structures are called resonance structures, and molecules such as benzene, which have two or more resonance structures, are said to exhibit resonance. The actual bonding in such molecules is thought to be an average of the bonding present in the resonance structures. The stability of molecules exhibiting resonance is found to be higher than that anticipated for any single resonance structure.

Although the conclusions we have drawn regarding molecular geometry and polarity can be obtained from Lewis structures, it is much easier to draw such conclusions from models of molecules and ions. The rules we have cited for octet rule structures transfer readily to models. In many ways the models are easier to construct than are the drawings of Lewis structures on paper. In addition, the models are three-dimensional and hence much more representative of the actual species. Using the models, it is relatively easy to see both geometry and polarity, as well as to deduce Lewis structures. In this experiment you will assemble models for a sizeable number of common chemical species and interpret them in the ways we have discussed.

Experimental Procedure

IN THIS EXPERIMENT YOUR INSTRUCTOR MAY ALLOW YOU TO WORK WITHOUT SAFETY GLASSES.

In this experiment you may work in pairs during the first portion of the laboratory period.

The models you will use consist of drilled wooden balls, short sticks, and springs. The balls represent atomic nuclei surrounded by the inner electron shells. The sticks and springs represent electron pairs and fit in the holes in the wooden balls. The model (molecule or ion) consists of wooden balls (atoms) connected by sticks or springs (chemical bonds). Some sticks may be connected to only one atom (nonbonding pairs).

In this experiment we will deal with atoms that obey the octet rule; such atoms have four electron pairs around the central core and will be represented by balls with four tetrahedral holes in which there are four sticks or springs. The only exception will be hydrogen atoms, which share two electrons in covalent compounds, and which will be represented by balls with a single hole in which there is a single stick.

In assembling a molecular model of the kind we are considering, it is possible, indeed desirable, to proceed in a systematic manner. We will illustrate the recommended procedure by developing a model for a molecule with the formula CH_2O.

1. Determine the total number of valence electrons in the species. This is easily done once you realize that the number of valence electrons on an atom is equal to the number of the group to which the atom belongs in the Periodic Table. For CH_2O,

    C   Group 4   H   Group 1   O   Group 6

    Therefore each carbon atom in a molecule or ion contributes four electrons, each hydrogen atom one electron, and each oxygen atom six electrons. The total number of valence electrons equals the sum of the valence electrons on all of the atoms in the species being studied. For CH_2O this total would be $4 + (2 \times 1) + 6$, or 12 valence electrons. If we are working with an ion, we add one electron for each negative charge or subtract one for each positive charge on the ion.

2. Select wooden balls and sticks to represent the atoms and electron pairs in the molecule. You should use four-holed balls for the carbon atom and the oxygen atom, and one-holed balls to represent the hydrogen atoms. Since there are 12 valence electrons in the molecule and electrons occur in pairs, you will need six sticks to represent the six electron pairs. The sticks will serve both as bonds between atoms and as nonbonding electron pairs.

3. Connect the balls with some of the sticks. (Assemble a skeleton structure for the molecule, joining atoms by single bonds.) In some cases this can only be done in one way. Usually, however, there are various possibilities, some of which are more reasonable than others. In CH_2O the model can be assembled by connecting the two H atom balls to the C atom ball with two of the available sticks, and then using a third stick to connect the C atom and O atom balls.

4. The next step is to use the sticks that are left over in such a way as to fill all the remaining holes in the balls. (Distribute the electron pairs so as to give each atom eight electrons and so satisfy the octet rule.) In the model we have assembled, there is one

unfilled hole in the C atom ball, three unfilled holes in the O atom ball, and three available sticks. An obvious way to meet the required condition is to use two sticks to fill two of the holes in the O atom ball, and then use two springs instead of two sticks to connect the C atom and O atom balls. The model as completed is shown in Figure 12.1.

5. Interpret the model in terms of the atoms and bonds represented. The sticks and spatial arrangement of the balls will closely correspond to the electronic and atomic arrangement in the molecule. Given our model, we would describe the CH_2O molecule as being planar with single bonds between carbon and hydrogen atoms and a double bond between the C and O atoms. The H—C—H angle is approximately tetrahedral. There are two nonbonding electron pairs on the O atom. Since all bonds are polar and the molecular symmetry does not cancel the polarity in CH_2O, the molecule is polar. The Lewis structure of the molecule is given below:

$$\ddot{\text{O}}{=}\text{C}\Big\langle{}^{\text{H}}_{\text{H}}$$

(The compound having molecules with the formula CH_2O is well known and is called formaldehyde. The bonding and structure in CH_2O are given by the model.)

6. Investigate the possibility of the existence of isomers or resonance structures. It turns out that in the case of CH_2O one can easily construct an isomeric form that obeys the octet rule, in which the central atom is oxygen rather than carbon. It is found that this isomeric form of CH_2O does not exist in nature. As a general rule, carbon atoms almost always form a total of four bonds; put another way, nonbonding electron pairs on carbon atoms are very rare. Another useful rule of a similar nature is that if a species contains several atoms of one kind and one of another, the atoms of the same kind will assume equivalent positions in the species. In SO_4^{2-}, for example, the four O atoms are all equivalent, and are bonded to the S atom and not to one another.

Resonance structures are reasonably common. For resonance to occur, however, the atomic arrangement must remain fixed for two or more possible electronic structures. For CH_2O there are no resonance structures.

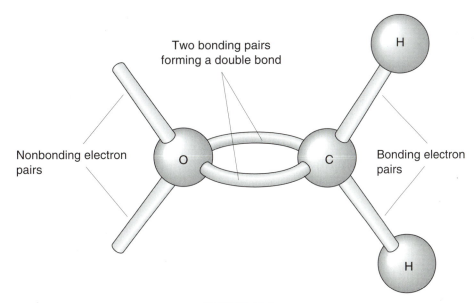

FIGURE 12.1

A. Using the procedure we have outlined, construct and report on models of the molecules and ions listed here and/or other species assigned by your instructor. Draw the complete Lewis structure for each molecule, showing nonbonding as well as bonding electrons. Given the structure, describe the geometry of the molecule or ion, and state whether the species is polar. Finally, draw the Lewis structures of any likely isomers or resonance forms.

CH_4	H_3O^+	N_2	C_2H_2	SCN^-
CH_2Cl_2	HF	P_4	SO_2	NO_3^-
CH_4O	NH_3	C_2H_4	SO_4^{2-}	HNO_3
H_2O	H_2O_2	$C_2H_2Br_2$	CO_2	$C_2H_4Cl_2$

B. Assuming that stability requires that each atom obey the octet rule, predict the stability of the following species:

$$PCl_3 \quad H_3O \quad CH_2 \quad CO$$

C. When you have completed parts A and B, see your laboratory instructor, who will check your results and assign you a set of unknown species. Working now by yourself, assemble models for each species as in the previous section, and report on the geometry and bonding in each of the unknown species on the basis of the model you construct. Also consider and report the polarity and the Lewis structures of any isomers and resonance forms for each species.

Name —————————————   *Section* ——————

Report: Geometrical Structure of Molecules Using Molecular Models

A.	*Species*	*Lewis structure*	*Molecular geometry*	*Polar?*	*Isomers or resonance structures*
	CH_4				
	CH_2Cl_2				
	CH_4O				
	H_2O				
	H_3O^+				
	HF				
	NH_3				
	H_2O_2				
	N_2				
	P_4				
	C_2H_4				

(continued on following page)

(continued)

Species	*Lewis structure*	*Molecular geometry*	*Polar?*	*Isomers or resonance structures*
$C_2H_2Br_2$				
C_2H_2				
SO_2				
SO_4^{2-}				
CO_2				
SCN^-				
NO_3^-				
HNO_3				
$C_2H_4Cl_2$				

B. Stability predicted for PCl_3 ________ H_3O ________ CH_2 ________ CO ________

C. ***Unknowns***

____

____

____

____

Advance Study Assignment: The Geometrical Structure of Molecules

You are asked by your instructor to construct a model of the CH_2Cl_2 molecule. Being of a conservative nature, you proceed as directed in the section on Experimental Procedure.

1. First you need to find the number of valence electrons in CH_2Cl_2. The number of valence electrons in an atom of an element is equal to the group number of that element in the Periodic Table.

    C is in Group _____ H is in Group _____ Cl is in Group _____

    In CH_2Cl_2 there is a total of _____ valence electrons.

2. The model consists of balls and sticks. What kind of ball should you select for the C atom? _____ the H atoms? _____ the Cl atom? _____ The electrons in the molecule are paired, and each stick represents an electron pair. How many sticks do you need? _____

3. Assemble a skeleton structure for the molecule, connecting the balls with sticks into one unit. Use the rule that C atoms form four bonds, whereas Cl atoms usually do not. Draw a sketch of the skeleton below:

4. How many sticks did you need to make the skeleton structure? _____ How many sticks are left over? _____ If your model is to obey the octet rule, each ball must have four sticks in it (except for hydrogen atom balls, which need only one). (Each atom in an octet rule species is surrounded by four pairs of electrons.) How many holes remain to be filled? _____ Fill them with the remaining sticks, which represent nonbonding electron pairs. Draw the complete Lewis structure for CH_2Cl_2 using lines for bonds and pairs of dots for nonbonding electrons.

5. Describe the geometry of the model, which is that of CH_2Cl_2. _____ Is the CH_2Cl_2 molecule polar? _____ Why?

    Would you expect CH_2Cl_2 to have any isomeric forms? _____ Explain your reasoning.

6. Would CH_2Cl_2 have any resonance structures? _____ If so, draw them below.

Experiment 13

HEAT EFFECTS AND CALORIMETRY

Heat is a form of energy, sometimes called thermal energy, that can pass spontaneously from an object at a high temperature to an object at a lower temperature. If the two objects are in contact, they will, given sufficient time, both reach the same temperature.

Heat flow is ordinarily measured in a device called a calorimeter. A calorimeter is simply a container with insulating walls, made so that essentially no heat is exchanged between the contents of the calorimeter and the surroundings. Within the calorimeter chemical reactions may occur or heat may pass from one part of the contents to another, but no heat flows into or out of the calorimeter from or to the surroundings.

A. SPECIFIC HEAT

When heat flows into a substance, the temperature of that substance will increase. The quantity of heat q required to cause a temperature change Δt of any substance is proportional to the mass m of the substance and the temperature change, as shown in Equation 1. The proportionality constant is called the specific heat, $S.H.$, of that substance.

$$q = (\text{specific heat}) \times m \times \Delta t = S.H. \times m \times \Delta t \tag{1}$$

The specific heat can be considered to be the amount of heat required to raise the temperature of one gram of the substance by 1°C (if you make m and Δt in Equation 1 both equal to 1, then q will equal $S.H.$). Amounts of heat are measured in either joules or calories. To raise the temperature of 1 g of water by 1°C, 4.18 joules of heat must be added to the water. The specific heat of water is therefore 4.18 joules/g°C. Since 4.18 joules equals 1 calorie, we can also say that the specific heat of water is 1 calorie/g°C. Ordinarily heat flow into or out of a substance is determined by the effect that that flow has on a known amount of water. Because water plays such an important role in these measurements, the calorie, which was the unit of heat most commonly used until recently, was actually defined to be equal to the specific heat of water.

The specific heat of a metal can readily be measured in a calorimeter. A weighed amount of metal is heated to some known temperature and is then quickly poured into a calorimeter that contains a measured amount of water at a known temperature. Heat flows from the metal to the water, and the two equilibrate at some temperature between the initial temperatures of the metal and the water.

Assuming that no heat is lost from the calorimeter to the surroundings, and that a negligible amount of heat is absorbed by the calorimeter walls, the amount of heat that flows from the metal as it cools is equal to the amount of heat absorbed by the water.

In thermodynamic terms, the heat flow for the metal is equal in magnitude but opposite in direction, and hence in sign, to that for the water. For the heat flow q,

$$q_{H_2O} = -q_{metal} \tag{2}$$

If we now express heat flow in terms of Equation 1 for both the water and the metal M, we get

$$q_{H_2O} = S.H._{H_2O} m_{H_2O} \Delta t_{H_2O} = -S.H._M m_M \Delta t_M \tag{3}$$

In this experiment we measure the masses of water and metal and their initial and final temperatures. (Note that $\Delta t_M < 0$ and $\Delta t_{H_2O} > 0$, since $\Delta t = t_{final} - t_{initial}$.) Given the specific heat of water, we can find the positive specific heat of the metal by Equation 3. We will use this procedure to obtain the specific heat of an unknown metal.

The specific heat of a metal is related in a simple way to its molar mass. Dulong and Petit discovered many years ago that about 25 joules were required to raise the temperature of one mole of many metals by 1°C. This relation, shown in Equation 4, is known as the Law of Dulong and Petit:

$$MM \cong \frac{25}{S.H. \ (\text{J/g°C})} \tag{4}$$

where MM is the molar mass of the metal. Once the specific heat of the metal is known, the approximate molar mass can be calculated by Equation 4. The Law of Dulong and Petit was one of the few rules available to early chemists in their studies of molar masses.

B. HEAT OF REACTION

When a chemical reaction occurs in water solution, the situation is similar to that which is present when a hot metal sample is put into water. With such a reaction there is an exchange of heat between the reaction mixture and the solvent, water. As in the specific heat experiment, the heat flow for the reaction mixture is equal in magnitude but opposite in sign to that for the water. The heat flow associated with the reaction mixture is also equal to the enthalpy change, ΔH, for the reaction, so we obtain the equation

$$q_{\text{reaction}} = \Delta H_{\text{reaction}} = -q_{H_2O} \tag{5}$$

By measuring the mass of the water used as solvent, and by observing the temperature change that the water undergoes, we can find q_{H_2O} by Equation 1 and ΔH by Equation 5. If the temperature of the water goes up, heat has been *given off* by the reaction mixture, so the reaction is *exo*thermic; q_{H_2O} is *positive* and ΔH is *negative*. If the temperature of the water goes down, the reaction mixture has *absorbed heat from* the water and the reaction is *endo*thermic. In this case q_{H_2O} is *negative* and ΔH is *positive*. Both exothermic and endothermic reactions are observed.

One of the simplest reactions that can be studied in solution occurs when a solid is dissolved in water. As an example of such a reaction note the solution of NaOH in water:

$$\text{NaOH(s)} \rightarrow \text{Na}^+(\text{aq}) + \text{OH}^-(\text{aq}); \quad \Delta H = \Delta H_{\text{solution}} \tag{6}$$

When this reaction occurs, the temperature of the solution becomes much higher than that of the NaOH and water that were used. If we dissolve a known amount of NaOH in a measured amount of water in a calorimeter, and measure the temperature change that occurs, we can use Equation 1 to find q_{H_2O} for the reaction and use Equation 5 to obtain ΔH. Noting that ΔH is directly proportional to the amount of NaOH used, we can easily calculate $\Delta H_{\text{solution}}$ for either a gram or a mole of NaOH. In the second part of this experiment you will measure $\Delta H_{\text{solution}}$ for an unknown ionic solid.

Chemical reactions often occur when solutions are mixed. A precipitate may form, in a reaction opposite in direction to that in Equation 6. A very common reaction is that of neutralization, which occurs when an acidic solution is mixed with one that is basic. In the last part of this experiment you will measure the heat effect when a solution of HCl, hydrochloric acid, is mixed with one containing NaOH, sodium hydroxide, which is basic. The heat effect is quite large, and is the result of the reaction between H^+ ions in the HCl solution with OH^- ions in the NaOH solution:

$$\text{H}^+(\text{aq}) + \text{OH}^-(\text{aq}) \rightarrow \text{H}_2\text{O} \quad \Delta H = \Delta H_{\text{neutralization}} \tag{7}$$

Experimental Procedure

A. Specific Heat

From the stockroom obtain a calorimeter, a sensitive thermometer, a sample of metal in a large stoppered test tube, and a sample of unknown solid. (The thermometer is very expensive, so be careful when handling it.)

The calorimeter consists of two nested expanded polystyrene coffee cups fitted with a styrofoam cover. There are two holes in the cover for a thermometer and a glass stirring rod with a loop bend on one end. Assemble the experimental setup as shown in Figure 13.1.

Fill a 400-cm³ beaker two-thirds full of water and begin heating it to boiling. While the water is heating, weigh your sample of unknown metal in the large stoppered test tube to the nearest 0.1 g on a top-loading or triple-beam balance. Pour the metal into a dry container and weigh the empty test tube and stopper. Replace the metal in the test tube and put the *loosely* stoppered tube into the hot water in the beaker. The water level in the beaker should be high enough so that the top of the metal is below the water surface. Continue heating the metal in the water for at least 10 minutes after the water begins to boil to ensure that the metal attains the temperature of the boiling water. Add water as necessary to maintain the water level.

While the water is boiling, weigh the calorimeter to 0.1 g. Place about 40 cm³ of water in the calorimeter and weigh again. Insert the stirrer and thermometer into the cover and put it on the calorimeter. The thermometer bulb should be completely under the water.

Measure the temperature of the water in the calorimeter to 0.1°C. Take the test tube out of the beaker of boiling water, remove the stopper, and pour the metal into the water in the calorimeter. Be careful that no water adhering to the outside of the test tube runs into the calorimeter when you are pouring the metal. Replace the calorimeter cover and agitate the water as best you can with the glass stirrer. Record to 0.1°C the maximum temperature

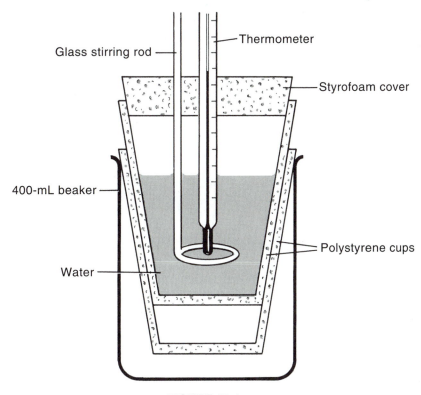

Glass stirring rod

Thermometer

Styrofoam cover

400-mL beaker

Polystyrene cups

Water

FIGURE 13.1

reached by the water. Repeat the experiment, using about 50 cm³ of water in the calorimeter. Be sure to dry your metal before reusing it; this can be done by heating the metal briefly in the test tube in boiling water and then pouring the metal onto a paper towel to drain. You can dry the hot test tube with a little compressed air.

B. Heat of Solution

Place about 50 cm³ of distilled water in the calorimeter and weigh as in the previous procedure. Measure the temperature of the water to 0.1°C. The temperature should be within a degree or two of room temperature. In a small beaker weigh out about 5 g of the solid compound assigned to you. Make the weighing of the beaker and of the beaker plus solid to 0.1 g. Add the compound to the calorimeter. Stirring continuously and occasionally swirling the calorimeter, determine to 0.1°C the maximum or minimum temperature reached as the solid dissolves. Check to make sure that *all* the solid dissolved. A temperature change of at least 5 degrees should be obtained in this experiment. If necessary, repeat the experiment, increasing the amount of solid used.

Dispose of the solution in Part B as directed by your instructor.

C. Heat of Neutralization

Rinse out your calorimeter with distilled water, pouring the rinse into the sink. In a graduated cylinder, measure out 25 cm³ of 1.00 M HCl; pour that solution into the calorimeter. Rinse out the cylinder with distilled water, and measure out 25 cm³ of 1.00 M NaOH; pour that solution into a dry 50-mL beaker. Measure the temperature of the acid and of the base to ± 0.1°C, making sure to rinse and dry your thermometer before immersing it in the solutions. Put the thermometer back in the calorimeter cover. Pour the NaOH solution into the HCl solution and put on the cover of the calorimeter. Stir the reaction mixture, and record the maximum temperature that is reached by the neutralized solution.

When you have completed the experiment, you may pour the neutralized solution down the sink. Return the calorimeter, thermometer, and metal sample to the stockroom.

Data and Calculations: Calorimetry

A. Specific Heat

	Trial 1	*Trial 2*
Mass of stoppered test tube plus metal	_____ g $\rightarrow$	_____ g
Mass of test tube and stopper	_____ g $\rightarrow$	_____ g
Mass of calorimeter	_____ g $\rightarrow$	_____ g
Mass of calorimeter and water	_____ g	_____ g
Mass of water	_____ g	_____ g
Mass of metal	_____ g $\rightarrow$	_____ g
Initial temperature of water in calorimeter	_____ °C	_____ °C
Initial temperature of metal (assume 100°C unless directed to do otherwise)	_____ °C $\rightarrow$	_____ °C
Equilibrium temperature of metal and water in calorimeter	_____ °C	_____ °C
Δt_{water} ($t_{final} - t_{initial}$)	_____ °C	_____ °C
Δt_{metal}	_____ °C	_____ °C
q_{H_2O}	_____ J	_____ J
Specific heat of the metal (Eq. 3)	_____ J/g°C	_____ J/g°C
Approximate molar mass of metal	_____	_____
Unknown no.	_____	

B. Heat of Solution

Mass of calorimeter plus water _____ g

Mass of beaker _____ g

(continued on following page)

Mass of beaker plus solid   _____ g

Mass of water, m_{H_2O} _____ g

Mass of solid, m_s _____ g

Original temperature   _____ °C

Final temperature   _____ °C

q_{H_2O} for the reaction (Eq. 1) (*S.H.* $= 4.18$ J/g°C)   _____ joules

ΔH for the reaction (Eq. 5)   _____ joules

The quantity you have just calculated is approximately* equal to the heat of solution of your sample. Calculate the heat of solution per gram of solid sample.

$$\Delta H_{solution} = \text{_____ joules/g}$$

The solution reaction is   endothermic   exothermic. (Underline correct answer.) Give your reasoning.

Solid unknown no. _____

(Optional) Formula of compound used (if furnished) _____ Molar mass _____ g

Heat of solution per mole of compound   _____ kJ

C. Heat of Neutralization

Original temperature of HCl solution   _____ °C

Original temperature of NaOH solution   _____ °C

Final temperature of neutralized mixture   _____ °C

Change in temperature, Δt (take average of the original temperatures of HCl and NaOH)   _____ °C

q_{H_2O} (assume 50 g H_2O are present)   _____ J

ΔH for the neutralization reaction   _____ J

ΔH per mole of H^+ and OH^- ions reacting   _____ kJ

---

* The value of ΔH will be approximate for several reasons. One is that we do not include the amount of heat absorbed by the solute. This effect is smaller than the likely experimental error, and thus we will ignore it.

Advance Study Assignment: Heat Effects and Calorimetry

1. A metal sample weighing 45.2 g and at a temperature of 100.0°C was placed in 38.6 g of water in a calorimeter at 25.2°C. At equilibrium the temperature of the water and metal was 33.0°C.

   a. What was Δt for the water? $(\Delta t = t_{final} - t_{initial})$

   _____ °C

   b. What was Δt for the metal?

   _____ °C

   c. How much heat flowed into the water?

   _____ joules

   d. Taking the specific heat of water to be 4.18 J/g°C, calculate the specific heat of the metal, using Equation 3.

   _____ joules/g°C

   e. What is the approximate molar mass of the metal? (Use Eq. 4.)

   _____ g

2. When 2.0 g of NaOH were dissolved in 49.0 g water in a calorimeter at 24.0°C, the temperature of the solution went up to 34.5°C.

   a. Is this solution reaction exothermic? _____ Why?

   b. Calculate q_{H_2O}, using Equation 1.

   _____ joules

   c. Find ΔH for the reaction as it occurred in the calorimeter (Eq. 5).

   $\Delta H =$ _____ joules

(continued on following page)

d. Find ΔH for the solution of 1.00 g NaOH in water.

$$\Delta H = \underline{\hspace{2cm}} \text{ joules/g}$$

e. Find ΔH for the solution of 1 mole NaOH in water.

$$\Delta H = \underline{\hspace{2cm}} \text{ joules/mole}$$

f. Given that NaOH exists as Na^+ and OH^- ions in solution, write the equation for the reaction that occurs when NaOH is dissolved in water.

g. Given the following heats of formation, ΔH_f, in kJ per mole, as obtained from a table of ΔH_f data, calculate ΔH for the reaction in Part f. Compare your answer with the result you obtained in Part e. NaOH(s), -426.7; Na^+, -239.7; OH^-, -229.9

$$\Delta H = \underline{\hspace{2cm}} \text{ kJ}$$

VAPOR PRESSURE AND HEAT OF VAPORIZATION OF LIQUIDS

The vapor pressure of a pure liquid is the total pressure at equilibrium in a container in which only the liquid and its vapor are present. In a container in which the liquid and other gas are both present, the vapor pressure of the liquid is equal to the partial pressure of its vapor in the container. In this experiment you will measure the vapor pressure of a liquid by determining the increase that occurs in the pressure in a closed container filled with air when the liquid is injected into it.

The vapor pressure of a liquid rises rapidly as the temperature is increased and reaches one atmosphere at the normal boiling point of the liquid. Thermodynamic arguments show that the vapor pressure of a liquid depends on temperature according to the equation:

$$\log_{10}VP = -\frac{\Delta H_{vap}}{2.3RT} + C \tag{1}$$

where VP is the vapor pressure, ΔH_{vap} is the amount of heat in joules required to vaporize one mole of the liquid against a constant pressure, R is the gas constant, 8.31 J/mole K, and T is the absolute temperature. You will note that this equation is of the form

$$Y = BX + C \tag{2}$$

where $Y = \log_{10}VP$, $X = 1/T$, and $B = -\Delta H_{vap}/2.3R$. Consequently, if we measure the vapor pressure of a liquid at various temperatures and plot $\log_{10}VP$ vs. $1/T$, we should obtain a straight line. From the slope B of this line, we can calculate the heat of vaporization of the liquid, since $\Delta H_{vap} = -2.3RB$.

In the laboratory you will measure the vapor pressure of an unknown liquid at approximately 0°C, 20°C, and 40°C, as well as its boiling point at atmospheric pressure. Given the three vapor pressures, you will be able to calculate the heat of vaporization of the liquid by making a graph of $\log_{10}VP$ vs. $1/T$. The graph will then be used to predict the boiling point of the liquid, and the value obtained will be compared with that you found experimentally.

Experimental Procedure

WEAR YOUR SAFETY GLASSES
WHILE PERFORMING THIS EXPERIMENT

From the stockroom obtain a suction flask, a rubber stopper fitted with a small dropper, and a short length of rubber tubing, all in a plastic bucket. Also obtain a sample of an unknown liquid.

1. Assemble the apparatus, using the mercury manometer at your lab bench, as indicated in Figure 14.1. The flask should be dry on the inside. If it is not, rinse it with a few milliliters of acetone (flammable) and blow compressed air into it for a few moments until it is dry. Put a piece of folded rectangular filter paper into the flask, so that liquid from the dropper will fall on the paper. This will facilitate the diffusion of the vapor. Pour some tap water into a beaker and bring it to about 20°C by adding some cold or warm water. Pour this water into the large beaker so that the level of water reaches as

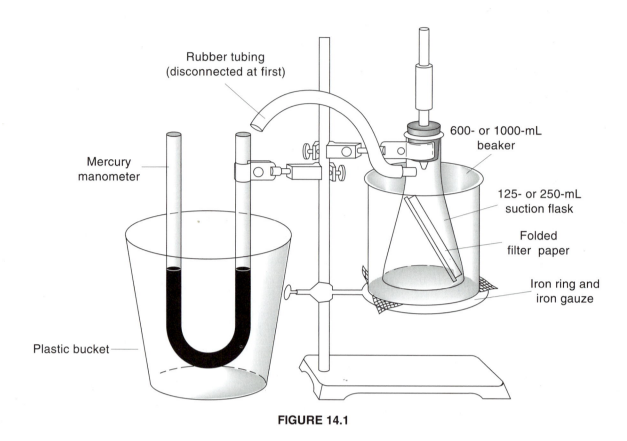

FIGURE 14.1

far as possible up the neck of the flask. Wait several minutes to ensure that the flask and air inside it are at the temperature of the water. Then remove the stopper from the flask. Pour a small amount of the unknown liquid into a small beaker, and draw about 2 mL of the liquid up into the dropper. Blot any excess liquid from the end of the dropper with a paper towel. Press the stopper *firmly* into the flask and connect the hose to the manometer. The mercury levels in the manometer should remain essentially equal.

Immediately squeeze the liquid from the dropper onto the paper, where it will vaporize, diffuse, and exert its vapor pressure. This vapor pressure will be equal to the increase in gas pressure at equilibrium in the container. If you do not observe any appreciable (>10 mm Hg) pressure increase within a minute or two after injecting the liquid, you probably have a leak in your apparatus and should consult your instructor. When the pressure in the flask becomes steady, in about 10 minutes, read and record the heights of the mercury levels in the manometer to ± 1 mm. An easy way to do this is to measure the height of each level above the lab bench top, using a ruler or meter stick. The *difference in height* is equal to the vapor pressure of the liquid in mm Hg at the temperature of the water bath. Record that temperature to $\pm 0.2°C$.

In this experiment it is essential that (1) the stopper is pressed firmly into the flask, so that the flask plus tubing plus manometer constitute a gas-tight system; (2) no liquid falls into the flask until the manometer is connected; (3) the operations of stoppering the flask, connecting the hose to the manometer, and squeezing the liquid into the flask are conducted with dispatch; (4) you take care when connecting the hose to the manometer, since mercury has a poisonous vapor and is not easily cleaned up when spilled.

Remove the flask from the system. Take out the filter paper. Dry the flask with compressed air, and dry the dropper by squeezing it several times in air.

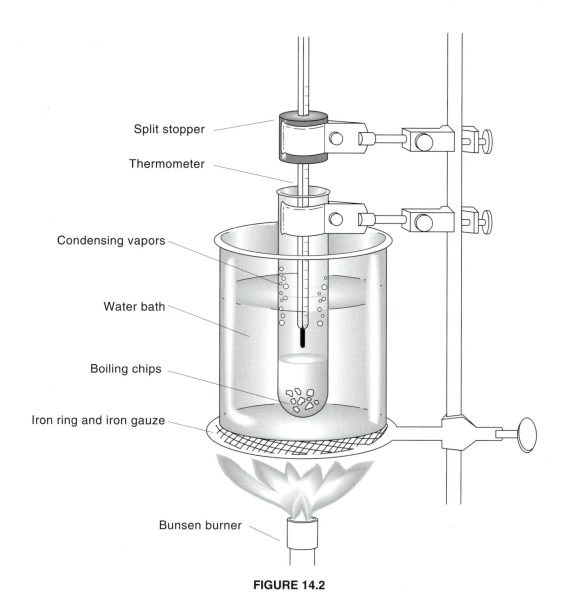

FIGURE 14.2

2. Starting again at 1, carry out the same experiment at about 0°C. Put a new piece of filter paper into the flask. Use an ice-water bath for cooling the flask. Measure the bath temperature, rather than assuming it to be 0°C.

3. Repeat the experiment once again, this time holding the water bath at 40°C by judicious heating with a Bunsen burner. Between each run, the flask and dropper must be thoroughly dried, and the precautions noted carefully observed.

4. In the last part of the experiment you will measure the boiling point of the liquid at the barometric pressure in the laboratory. This is done by pouring the remaining sample of unknown into a large test tube. Determine the boiling point of the liquid using the apparatus shown in Figure 14.2. The thermometer bulb should be just above the liquid surface. Heat the water bath until the liquid in the tube boils gently, with its vapor condensing *at least 5 cm below* the top of the test tube. Boiling chips may help in keeping the liquid boiling smoothly. As the boiling proceeds there will be some condensation on the thermometer and droplets will be falling from the thermometer bulb. After a minute or two the temperature should become reasonably steady at the boiling point of the liquid. Record the temperature under these conditions, along with the barometric pressure in the laboratory. The liquids used may be flammable and toxic, so

you should not inhale their vapors unnecessarily. *Do not* heat the water bath so strongly that condensation of the vapors from the liquid occurs only at the top of the test tube.

When you have completed this part of the experiment, pour the liquid remaining in the test tube into the waste crock.

C A U T I O N: *Mercury vapor is toxic. Use care when working with the mercury manometer. Keep the plastic bucket under the manometer to catch the mercury in case of a spill.*

Data and Calculations: Vapor Pressure and Heat of Vaporization of Liquids

	Temperature, t, *in °C*	*Heights of manometer* *mercury levels in mm*		*Vapor pressure* *mm Hg*
1.	_____	_____	_____	_____
2.	_____	_____	_____	_____
3.	_____	_____	_____	_____

Boiling point _____ °C

Barometric pressure _____ mm Hg

Using the relation (1) between vapor pressure and temperature, we will calculate the molar heat of vaporization of your liquid and its boiling point from the vapor pressure data obtained. It would be useful to first make the calculations indicated in the following table:

Approximate temperature °C	*t, actual temperature* °C	*T, temperature* K	*1/T*	*Vapor pressure, VP, in mm Hg*	*log$_{10}$VP*
0	_____	_____	_____	_____	_____
20	_____	_____	_____	_____	_____
40	_____	_____	_____	_____	_____

On the graph paper provided make a graph of $\log_{10}VP$ vs. $1/T$. Let $\log_{10}VP$ be the ordinate and plot $1/T$ on the abscissa. Since $\log_{10}VP$ is, by (1), a linear function of $1/T$, the line obtained should be nearly straight. Find the slope of the line, $\Delta\log_{10}VP/\Delta(1/T)$. (See Appendix V.)

The slope of the line is equal to $-\Delta H_{vaporization}/2.3R$. Given that $R = 8.31$ joules/mole K, calculate the molar heat of vaporization, ΔH_{vap}, of your liquid.

$$\text{Slope} = \frac{\Delta\log_{10}VP}{\Delta 1/T} = \text{_____} = \frac{-\Delta H_{vap}}{2.3R}, \quad \Delta H_{vap} = \text{_____ joules/mole}$$

From the graph it is also possible to find the temperature at which your liquid will have any given vapor pressure. Recalling that a liquid will boil in an open container at the *temperature* at which its *vapor pressure* is equal to the *atmospheric pressure,* predict the boiling point of the liquid at the barometric pressure in the laboratory.

(continued on following page)

(continued)

$P_{barometric}$ _____ mm Hg

$\log_{10} P_{barometric}$ _____

$1/T$ at this pressure   _____ K^{-1} (from graph)

T at this pressure   _____ K

t at this pressure   _____ °C (boiling point predicted)

Boiling point observed experimentally _____ °C

Unknown no. _____

Do your data appear to follow the equation:

$$\log_{10} VP = \frac{-\Delta H_{vap}}{2.3RT} + C$$

Give your reasoning.

Name _____ Section _____

VAPOR PRESSURE AND HEAT OF
VAPORIZATION OF LIQUIDS
(Data and Calculations)

Advance Study Assignment: Vapor Pressure of Liquids

The vapor pressure of ethyl alcohol was measured by the method of this experiment with the following results (see Fig. 14.1):

$t°C$	Heights of manometer levels in mm Hg Right	Left	Vapor pressure in mm Hg	$log_{10}VP$	T in K	$1/T$
− 2	84.5	94.5	_____	_____	_____	_____
19	69.5	109.5	_____	_____	_____	_____
35	39.5	139.5	_____	_____	_____	_____

   a. Calculate the vapor pressure of ethyl alcohol at the three temperatures. Find the logarithm to base 10 of each vapor pressure. Enter your results in the table.
   b. Find the absolute temperature T and its reciprocal at each temperature.
   c. Using the graph paper on the next page, plot the three points relating $log_{10}VP$ to $1/T$. Draw the best straight line you can through these points. (Such a line minimizes the sum of the distances from the points to the line.) See Appendix V.
   d. Determine the slope of the straight line. To do this you need to find $\Delta log_{10}VP/\Delta(1/T)$. Perhaps the easiest way to evaluate these terms is to extend the straight line until it cuts the ordinate (the vertical axis). Take a section of the straight line, say from $1/T$ equals 0.00275 to 0.00375. For that section, $\Delta 1/T$ equals $0.00375 − 0.00275$, or 0.00100. $\Delta log_{10}VP$ equals the change in $log_{10}VP$ when you go from the left end of the line to the right ($\Delta log_{10}VP$ is negative). To obtain the slope, divide $\Delta log_{10}VP$ by $\Delta(1/T)$.

slope = _____

   e. Noting that $\Delta H_{vap} = − 2.3R \times$ slope, calculate the molar heat of vaporization of ethyl alcohol. (Heats of vaporization of typical liquids are about 30,000 joules/mole.)

$\Delta H_{vap} =$ _____ joules/mole

   f. Ethyl alcohol will boil when its vapor pressure exceeds the air pressure at its surface. Using the graph from Part c, find $1/T$ when the vapor pressure is 760 mm Hg. Then find T in K and in °C. This temperature is the normal boiling point of ethyl alcohol.

$1/T$ _____: $T =$ _____ K: $t =$ _____°C: $t_{obs} = 78.4°C$

VAPOR PRESSURE OF LIQUIDS
(Advance Study Assignment)

xperiment 15

THE STRUCTURE OF CRYSTALS—
AN EXPERIMENT USING MODELS

If one examines the crystals of an ordinary substance like table salt, using a magnifying glass or a microscope, one finds many cubic particles, among others, in which planes at right angles are present. This is the situation with many common solids. The regularity we see implies a deeper regularity in the arrangement of atoms or ions in the solid. Indeed, when we study crystals by x-ray diffraction, we find that the atomic nuclei are present in remarkably symmetrical arrays, which continue in three dimensions for thousands or millions of units. Substances having a regular arrangement of atom-size particles in the solid are called crystalline, and the solid material consists of crystals. This experiment deals with some of the simpler arrays in which atoms or ions occur in crystals, and what these arrays can tell us about such properties as atomic sizes, densities of solids, and the efficiency of packing of particles.

Many crystals are complex, almost beyond belief. We will limit ourselves to the simplest crystals, those which have cubic structures. They are by far the easiest to understand, yet they exhibit many of the interesting properties of more complicated structures. We will further limit our discussion to substances containing only one or two kinds of atoms, so we will be working with the crystals of some of the elements and some binary compounds.

The atoms in crystals occur in planes. Sometimes a crystal will cleave readily at such planes. This is the reason the cubic structure of salt is so apparent. Let us begin our study of crystals with a simple example, a two-dimensional crystal in which all of the atoms lie in a square array in a plane, in the manner shown below:

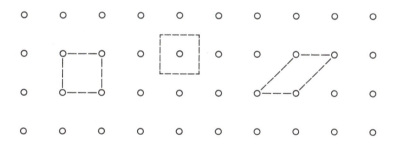

The first thing that you need to realize is that the array goes on essentially forever, in both directions. In this array ○ is always the same kind of atom, so all the atoms are equivalent. The distance, d_o, between atoms in the vertical direction is the same as in the horizontal, so if you knew either distance you could generate the array completely. In crystal studies we select a small section of the array, called the unit cell, that represents the array in the sense that by moving the unit cell repeatedly in either the x or y direction, a distance d_o at a time, you could generate the entire array. In the sketch we have used dotted lines to indicate several possible unit cells. All the cells have the same area, and by moving them up or down or across we could locate all the sites in the array. Ordinarily we select the unit cell on the left, because it includes four of the atoms and has its edges along the natural axes, x and y, for the array, but that is not necessary. In fact, the middle cell has the advantage that it clearly tells us that the number of ○ atoms in the cell is equal to 1. The number of atoms in whatever cell we choose must also equal 1, but that is not so apparent in the cell on the left. However, once you realize that only 1/4 of each atom on the corners of the cell actually belongs to that cell, because it is shared by three other cells, the number of atoms in the whole cell becomes equal to 1/4 × 4, or 1, which is what we got by drawing the cell in a different position.

When we extend the array to three dimensions, the same ideas regarding unit cells apply. The unit cell is the smallest portion of the array that could be used to generate the array. With cubic cells the unit cell is usually chosen to have edges parallel to the x, y, and z axes we could put on the array.

Experimental Procedure

IN THIS EXPERIMENT YOUR INSTRUCTOR MAY ALLOW YOU TO WORK WITHOUT SAFETY GLASSES.

In this experiment we will mix the discussion with the experimental procedure, since one supports and illustrates the other. We will start with the simplest possible cubic crystal, deal with its properties, and then go on to the next more complex example, and the next, and so on. Each kind of crystal will be related in some ways to the earlier ones but will have its own properties as well. So get out your model set and let's begin.

Work in pairs, and complete each section before going on to the next, unless directed otherwise by your instructor.

The Simple Cubic Crystal

You can probably guess the form of the array in the crystal structure we call simple cubic (SC). It is shown on the Calculations page at the end of this section.

The unit cell is a cube with an edge length equal to the distance from the center of one atom to the center of the next. The cell edge is usually given the symbol d_o. The volume of the unit cell is d_o^3, and is very small, since d_o is of the order of 0.5 nm. Using x-rays we can measure d_o to four significant figures quite easily. The number of atoms in the unit cell in a simple cubic crystal is equal to 1. Can you see why? Only 1/8 of each corner atom actually belongs to the cell, since it is shared equally by eight cells.

Using your model set, assemble three attached unit cells having the simple cubic structure shown in the sketch. Use the short bonds (no. 6) and the gray atoms. Each bond goes into a square face on the atom. An actual crystal with this structure would have many such cells, in three dimensions.

If you extended the model you have made further, you would find that each atom would be connected to six others. We say that the coordination number of the atoms in this structure is 6. Only the closest atoms are considered to be bonded to each other. In the other models we will be making, *only* those atoms that are *bonded*, by covalent or ionic bonds, will be connected to one another, so the number of bonds to an atom in a large model will be equal to the coordination number.

Although the model has an open structure to help us see relationships better, in an actual crystal we consider that the atoms that are closest are touching. It is on this assumption that we determine atomic radii. In this SC crystal, if we know d_o we can find the atomic radius r of the atoms, since, if the atoms are touching, d_o must equal $2r$. Another property we can calculate knowing d_o is the density, given the nature of the atoms in the crystal. From d_o we can easily find the volume V of the unit cell. Since there is one atom per cell, a mole will contain Avogadro's number of cells. Given the molar mass of the element, we can find the mass m of a cell. The density is simply m/V. With more complex crystals we can make these same calculations, but must take account of the fact that the number of atoms or ions per unit cell may not be equal to one.

Essentially no elements crystallize in an SC structure. The reason is that SC packing is inefficient, in that the atoms are farther apart than they need be. There is, as you can see, a big hole in the middle of each unit cell that is begging for an atom to go there, and atoms indeed do. Before we go into that, let's calculate the fraction of the volume of the unit cell that is actually occupied by atoms. This is easy to do. Make the calculation of that fraction on the Calculations page.

That fraction is indeed pretty small; only about 52 percent of the cell volume is occupied by atoms. Most of the empty space is in that hole. You can calculate by simple geometry that an atom having a radius equal to 73 percent of that of the atom on a corner would fit into the hole. Or, putting it another way, an atom bigger than that would have to push the corner atoms back to fit in. If the particles on the corners were anions, and the atom in the hole were a cation, the cation in the hole would push the anions apart, decreasing the repulsion between them and maximizing the attraction between anions and cations. We will have more to say about this when we deal with binary salts, but for now you need to note that if $r_+/r_- > 0.732$, the cation will not fit in the cubic hole formed by anions on the corners of an SC cell.

Body-Centered Cubic Crystals

In a body-centered cubic (BCC) crystal, the unit cell still contains the corner atoms present in the SC structure, but in the center of the cell there is another atom of the same kind. The unit cell is shown on the Calculations page.

Using your model set, assemble a BCC crystal. Use the blue balls. Put the short bonds in the eight holes in the triangular faces on one blue ball. Attach a blue ball to each bond, again using the holes in a triangular face. Those eight blue balls define the unit cell in this structure. Add as many atoms to the structure as you have available, so that you can see how the atoms are arranged. In a BCC crystal each atom is bonded to eight others, so the coordination number is 8. Verify this with your model. (There are no bonds along the edges of the unit cell, since the corner atoms are not as close to one another as they are to the central atom.)

The BCC lattice is much more stable than the SC one, in part at least because of the higher coordination number. Many metals crystallize in a BCC lattice—including sodium, chromium, tungsten, and iron—when these metals are at room temperature.

There are several properties of the BCC structure that you should note. The number of atoms per unit cell is two, one from the corner atoms and one from the atom in the middle of the cell, where it is unshared. As with SC cells, there is a relation between the unit cell size and the atom radius. Given that in sodium metal the cell edge is 0.429 nm, calculate the radius of a sodium atom on the Calculations page. When you are finished, calculate the density of sodium metal.

The fraction of the volume that is occupied by atoms in a BCC crystal is quite a bit larger than with SC crystals. Using the same kind of procedure that we did with the SC structure, we can show that in BCC crystals about 68 percent of the cell volume is occupied.

Close-Packed Structures

Although many elements form BCC crystals, still more prefer structures in which the atoms are close packed. In such structures there are layers of atoms in which each atom is in contact with six others, as in the sketch below:

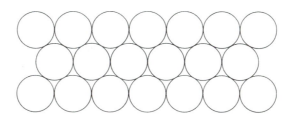

This is the way that billiard balls lie in a rack, or the honeycomb cells are arranged in a bees' nest. It is the most efficient way one can pack spheres, with about 74 percent of the volume in a close-packed structure filled with atoms. It turns out that there is more than one close-packed crystal structure. The layers all have the same structure, but they can be stacked on one another in two different ways. This is certainly not apparent at first sight. The first close-packed crystal we will examine is cubic, amazingly enough, considering all those triangles in the layers of atoms.

Face-Centered Cubic Crystals

In the face-centered cubic (FCC) unit cell there are atoms on the corners and an atom at the center of each face. There is no atom in the center of the cell (see the sketch on the Calculations page, where we show the bonding on only the three exposed faces).

Constructing an FCC unit cell, using bonds between closest atoms, is not as simple as you might think, so let's do it the easy way. This time select the large gray balls, the ones with the most holes. Make the bottom face of the unit cell by attaching four gray balls to a center ball, using short bonds, and holes that are in rectangular faces. You should get the structure on the left below:

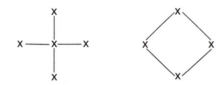

To make the middle layer of atoms, connect four of the balls in a square, again using short bonds and the holes in rectangular faces, as in the right sketch. The top layer is made the same way as the bottom one. Then connect the layers to one another, again using holes in the rectangular faces. You should then have an FCC cubic unit cell. Because the atoms in the middle of adjacent faces are as close to one another as they are to the corner atoms, there should be bonds between them. There are quite a few bonds in the final cell, 32 in all. Put them all in. As with BCC there are no bonds along the edges of the cell.

The FCC, or cubic close-packed, structure is a common one. Among the metals with this type of crystal are copper, silver, nickel, and calcium.

In close-packed structures the coordination number is the largest that is possible. It can be found by looking at a face-centered atom, say on the top face of the cell. That atom has eight bonds to it and would also have bonds to atoms in the cell immediately above, four of them. So its coordination number, like that of every atom in the cell, is 12.

Having seen how to deal with other structures, you should now be able to find the number of atoms in the FCC unit cell and the radius of an atom as related to d_o. From these relationships and the measured density, you can calculate Avogadro's number. On the Calculations page, find these quantities, using copper metal, which has a unit cell edge d_o equal to 0.361 nm and a density equal to 8.92 g/cm^3.

In the center of the unit cell there is a hole. It is smaller than a cubic hole but of significant size nonetheless. It is called an octahedral hole, because the six atoms around it define an octahedron. In an ionic crystal, the anions often occupy the sites of the atoms in your cell, and there is a cation in the center of the octahedral hole. On the Calculations page, find the maximum radius, r_+, of the cation that would just fit in an octahedral hole surrounded by anions of radius r_-. You should find that the r_+/r_- ratio turns out to be 0.414.

There is another kind of hole in the FCC lattice that is important. See if you can find it; there are eight of them in the unit cell. If you look at an atom on the corner of the cell, you can see that it lies in a tetrahedron. In the center of the tetrahedron there is a tetrahedral hole, which is small compared to an octahedral or cubic hole. However, in some crystals small cations are found in some or all of the tetrahedral holes, again in a close-packed anion lattice. The cation-anion radius ratio at which cations would just fit into a tetrahedral hole is not so easy to find; r_+/r_- turns out to be 0.225.

The close-packed layers of atoms in the FCC structure are not parallel to the unit cell faces, but rather are perpendicular to the cell diagonal. If you look down the cell diagonal, you see six atoms in a close-packed triangle in the layer immediately behind the corner atom, and another layer of close-packed atoms below that, followed by another corner atom. The layers are indeed closely packed, and, as one goes down the diagonal of this and succeeding cells, the layers repeat their positions in the order ABCABC . . . , meaning that atoms in every fourth layer lie below one another.

Clearly there is another way we could stack the layers. The first and second will always be in the same relative positions, but the third layer could be below the first one if it were shifted properly. So we can have a close-packed structure in which the order of the layers is ABA-BAB. . . . The crystal obtained from this arrangement of layers is not cubic, but hexagonal. It too is a common structure for metals. Cadmium, zinc, and manganese have this structure. As you might expect, the stability of this structure is very similar to that of FCC crystals. We find that simply changing the temperature often converts a metal from one form to another. Calcium, for example, is FCC at room temperature, but if heated to 450°C it converts to close-packed hexagonal.

Crystal Structures of Some Common Binary Compounds

We have now dealt with all of the possible cubic crystal structures for metals. It turns out that the structures of binary ionic compounds are often related to these metal structures in a very simple way. In many ionic crystals the anions, which are large compared with cations, are essentially in contact with each other, in either an SC or FCC structure. The cations go into the cubic, or octahedral, or tetrahedral holes, depending on the cation-anion radius ratios that we calculated. The idea is that the cation will tend to go into a hole in which it will not quite fit. This increases the unit cell size from the value it would have if the anions were touching, which reduces the repulsion energy due to anion-anion interaction and increases to the maximum the cation-anion attraction energy, producing the most stable possible crystal structure. According to the so-called radius-ratio rule, large cations go into cubic holes, smaller ones into octahedral holes, and the smallest ones into tetrahedral holes.

The deciding factor for which hole is favored is given by the radius ratio:

$$\text{If } r_+/r_- > 0.732 \qquad \text{cations go into cubic holes}$$
$$\text{If } 0.732 > r_+/r_- > 0.414 \quad \text{cations go into octahedral holes}$$
$$\text{If } 0.414 > r_+/r_- > 0.225 \quad \text{cations go into tetrahedral holes}$$

The NaCl Crystal

To apply the radius ratio rule to NaCl, we simply need to find r_+/r_-, using the data in Table 15.1. Since $r_{Na^+} = 0.095$ nm, and $r_{Cl^-} = 0.181$ nm, the radius ratio is 0.095/0.181, or 0.525. That value is less than 0.732 and greater than 0.414, so the sodium ions should go into octahedral holes. We saw the octahedral hole in the center of the FCC unit cell, so Na^+ ions go there, and the Cl^- ions have the FCC structure. Actually, there are 12 other octahedral holes associated with the cell, one on each edge, which would be apparent if we had been able to make more cells. An Na^+ ion goes into each of these holes, giving the classic NaCl structure shown on the Calculations page. (Again, only the ions and bonds on the exposed faces are shown.)

Make a model of the unit cell for NaCl, using the large gray balls for the Cl⁻ ions and the blue ones for Na⁺ ions. Clearly the Na⁺ ion at the center of the cell is in an octahedral hole, but so are all of the other sodium ions, because if you extend the lattice, every Na⁺ ion will be surrounded by six Cl⁻ ions. The coordination number of Na⁺, and of Cl⁻, ions is 6. The crystal is FCC in Cl⁻ and also in Na⁺, because you could put Na⁺ ions on the corners of the unit cell and maintain the same structure. The unit cell extends from the center of one Cl⁻ ion to the center of the next Cl⁻ along the cell edge; or, from the center of one Na⁺ ion to the center of the next Na⁺ ion.

On the Calculations page find the number of Na⁺ and of Cl⁻ ions in the unit cell.

The CsCl Crystal

Cesium chloride has the same type of formula as NaCl, 1 : 1. The Cs⁺ ion, however, is larger than Na⁺, and has a radius equal to 0.169 nm. This makes r_+/r_- equal to 0.933. Because this value is greater than 0.732, we would expect that in the CsCl crystal the Cs⁺ ions will fill cubic holes, and this is what is observed. The structure of CsCl will look like that of the BCC unit cell you made earlier, except that the ion in the center will be Cs⁺ and those on the corners Cl⁻. If you put gray balls in the center of each BCC unit cell you made from the blue balls, you would have the CsCl structure. This structure is *not* BCC, because the corner and center atoms are not the same. Rather it consists of two interpenetrating simple cubic lattices, one made from Cl⁻ ions and the other from Cs⁺ ions.

The Zinc Sulfide Crystal

Zinc sulfide is another 1 : 1 compound, but its crystal structure is not that of NaCl or of CsCl. In ZnS, the Zn²⁺ ions have a radius of 0.074 nm and the S²⁻ ions a radius of 0.184 nm, making r_+/r_- equal to 0.402. By the radius-ratio rule, ZnS should have close-packed S²⁻ ions, with the Zn²⁺ ions in tetrahedral holes. In the ZnS unit cell, we find that this is indeed the case; the S²⁻ ions are FCC, and alternate tetrahedral holes in the unit cell are occupied by Zn²⁺ ions, which themselves form a tetrahedron.

To make a model for ZnS, first assemble an SC unit cell, using blue balls and the long bonds. Use gray balls for the zinc ions, and assemble the unit shown below, using the short bonds and the holes in triangular faces:

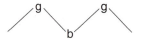

Attach this unit to two corner atoms on the diagonal of the bottom face. Make another unit and attach it to the two corner atoms in the upper face that lie on the face diagonal that is not parallel to the bottom face diagonal. The gray balls should then form a tetrahedron. Then attach the four other FCC blue balls to the gray ones. In this structure the coordination number is four (the long bonds do not count, they just keep the unit cell from falling apart). If the gray balls in the unit cell are replaced with blue ones, so that all atoms are of the same element, we obtain the diamond crystal structure.

These are the three common cubic structures of 1 : 1 compounds. The radius-ratio rule allows us to predict which structure a given compound will have. It does not always work, but it is correct most of the time. On the Calculations page, use the rule to predict the cubic structures that crystals of the following substances will have: KI, CuBr, and TlBr.

The Calcium Fluoride Crystal

Calcium fluoride is a 1 : 2 compound, so it cannot have the structure of any crystal we have discussed so far. The radii of Ca²⁺ and F⁻ are 0.099 and 0.136 nm, respectively, so r_+/r_- is

0.727. This makes CaF_2 on the boundary between compounds with cations in cubic holes or octahedral holes. It turns out that in CaF_2, the F^- ions have a simple cubic structure, with half of the cubic holes filled by Ca^{2+} ions. This produces a crystal in which the Ca^{2+} ions lie in an FCC lattice, with 8 F^- ions in a cube inside the unit cell.

Use your model set to make a unit cell for CaF_2. Use gray balls for the Ca^{2+} ions and red ones for F^-. The cations and anions are linked by short bonds that go into holes in nonadjacent triangular faces. The Ca^{2+} ions on each face diagonal are linked to F^- ions as shown below:

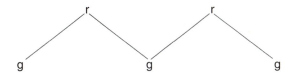

To complete the bottom face, attach two red and two gray balls to the initial line, so that the four red balls form a square and the gray ones an FCC face. The top of the unit cell is made the same way. Attach the two assemblies through four gray balls, which lie at the centers of the other faces. When you are done, the red balls should form a cube inside an FCC unit cell made from gray balls. Extend the model into an adjacent unit cell to show that every other cube of F^- ions has a Ca^{2+} ion at its center.

When you have completed this part of the experiment, your instructor may assign you a model with which to work. Report your results as directed. Then disassemble the models you made and pack the components in the box.

Calculations: The Structure of Crystals

Table 15.1

Atom	Molar Mass, g/mole	Atomic Radius, nm	Ionic Radius, nm
Na	22.99	—	0.095
Cu	63.55	—	0.096
Cl	35.45	0.099	0.181
Cs	132.9	0.262	0.169
Zn	65.38	0.133	0.074
S	32.06	0.104	0.184
K	39.10	0.231	0.133
I	126.9	0.133	0.216
Br	79.90	0.114	0.195
Tl	204.4	0.171	0.147

Some Cubic Unit Cells:

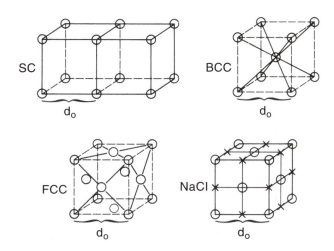

Simple Cubic Crystal

Fraction of volume of unit cell occupied by atoms

Volume of unit cell in terms of d_o _____

Number of atoms per unit cell   _____

Radius r of atom in terms of d_o _____

Volume of atom in terms of d_o $\left(V = \dfrac{4\pi}{3}\, r^3 \right)$ _____

Volume of atom/volume of cell   _____ = _____

(continued on following page)

Body-Centered Cubic

a. Radius of a sodium atom

In BCC atoms touch along the cube diagonal
(see your model)

Length of cube diagonal if $d_o = 0.429$ nm _____ nm

$4 \times r_{Na}$ = length of cube diagonal $\quad r_{Na} =$ _____ nm

b. Density of sodium metal

Length of unit cell, d_o, in cm (1 cm $= 10^7$ nm) _____ cm

Volume of unit cell, V _____ cm^3

Number of atoms per unit cell _____

Number of unit cells per mole Na _____

Mass of a unit cell, m _____ g

Density of sodium metal, m/V (Obs. 0.97 g/cm^3) _____ g/cm^3

Face-Centered Cubic

a. Number of atoms per unit cell

Number of atoms on corners _____ Shared by _____ cells _____ $\times$ _____ = _____

Number of atoms on faces _____ Shared by _____ cells _____ $\times$ _____ = _____

Total atoms per cell _____

b. Radius of a Cu atom

In FCC atoms touch along face diagonal (see your model and ASA)

Length of face diagonal in Cu, where $d_o = 0.361$ nm _____ nm

Number of Cu atom radii on face diagonal _____ $\quad r_{Cu} =$ _____ nm

(continued on following page)

c. Avogadro's Number, N

Length of cell edge, d_o, in cm   _____ cm

Volume of a unit cell in Cu metal   _____ cm³

Volume of a mole of Cu metal (V = mass/density)   _____ cm³

Number of unit cells per mole Cu   _____

Number of atoms per unit cell   _____

Number of atoms per mole, N _____

d. Size of an octahedral hole

Below is a sketch of a cross section of a unit cell, taken halfway between the top and bottom faces. The atoms are drawn touching, as they would be in the crystal. The cation goes in the hole in the middle.

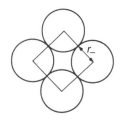

Show the length of d_o on the sketch (see your model).

What is the relationship between r_- and d_o?

$$r_- = \text{_____} \times d_o$$

What is the equation relating the length of the diagonal of the square to d_o (see your model)?

$$\text{diagonal} = \text{_____} \times d_o$$

What is the equation relating r_- and r_+ to the diagonal length? to d_o? Look at the sketch, and you should see the relation easily.

What is r_+ in terms of d_o?

$$r_+ = \text{_____} \times d_o$$

(continued on following page)

(continued)

What is the value of the radius ratio, r_+/r_-?

$$r_+/r_- = \underline{\hspace{3cm}}$$

NaCl

Number of Cl^- ions in the unit cell (FCC) $\underline{\hspace{3cm}}$

Number of Na^+ ions on edges of cell $\underline{\hspace{2cm}}$ Shared by $\underline{\hspace{2cm}}$ cells

Number of Na^+ ions in center of cell $\underline{\hspace{2cm}}$ Shared by $\underline{\hspace{2cm}}$ cell(s)

Total no. Na^+ ions in unit cell $\underline{\hspace{3cm}}$

Radius ratio rule (data in Table 15.1)

KI $\quad r_+/r_-$ $\underline{\hspace{2cm}}$ Predicted structure $\underline{\hspace{2cm}}$ Obs. NaCl

CuBr $\quad r_+/r_-$ $\underline{\hspace{2cm}}$ Predicted structure $\underline{\hspace{2cm}}$ Obs. ZnS

TlBr $\quad r_+/r_-$ $\underline{\hspace{2cm}}$ Predicted structure $\underline{\hspace{2cm}}$ Obs. CsCl

Report on unknown

Advance Study Assignment: Structure of Crystals

1. Many substances crystallize in a cubic structure. The unit cell for such crystals is a cube having an edge with a length equal to d_o.

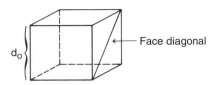

a. What is the length, in terms of d_o, of the face diagonal, which runs diagonally across one face of the cube?

_____

b. What is the length, again in terms of d_o, of the cube diagonal, which runs from one corner, through the center of the cube, to the other corner? Hint: Make a right triangle having a face diagonal and an edge of the cube as its sides, with the hypotenuse equal to the cube diagonal.

_____

2. In an FCC structure, the atoms are found on the corners of the cubic unit cell and in the centers of each face. The unit cell has an edge whose length is the distance from the center of one corner atom to the center of another corner atom on the same edge. The atoms on the diagonal of any face are touching. One of the faces of the unit cell is shown below.

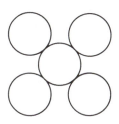

a. Show the distance d_o on the sketch. Draw the boundaries of the unit cell.

b. What is the relationship between the length of the face diagonal and the radius of the atoms in the cell?

Face diagonal = _____

(continued on following page)

(continued)

c. How is the radius of the atoms related to $d_\circ$?

$$r = \text{\underline{\hspace{3cm}}}$$

d. Silver metal crystals have an FCC structure. The unit cell edge in silver is 0.4086 nm long. What is the radius of a silver atom?

\text{\underline{\hspace{3cm}}} nm

Experiment 16

CLASSIFICATION OF CHEMICAL SUBSTANCES

Depending on the kind of bonding present in a chemical substance, the substance may be called ionic, molecular, or metallic.

In a solid ionic compound there are ions; the large electrostatic forces between the positively and negatively charged ions are responsible for the bonding which holds these particles together.

In a molecular substance the bonding is caused by the sharing of electrons by atoms. When the stable aggregates resulting from covalent bonding contain relatively small numbers of atoms, they are called molecules. If the aggregates are very large and include essentially all the atoms in a macroscopic particle, the substance is called macromolecular.

Metals are characterized by a kind of bonding in which the electrons are much freer to move than in other kinds of substances. The metallic bond is stable but is probably less localized than other bonds.

The terms ionic, molecular, macromolecular, and metallic are somewhat arbitrary, and some substances have properties that would place them in a borderline category, somewhere between one group and another. It is useful, however, to consider some of the general characteristics of typical ionic, molecular and macromolecular, and metallic substances, since many very common substances can be readily assigned to one category or another.

IONIC SUBSTANCES

Ionic substances are all solids at room temperature. They are typically crystalline but may exist as fine powders as well as clearly defined crystals. While many ionic substances are stable up to their melting points, some decompose on heating. It is very common for an ionic crystal to release loosely bound water of hydration at temperatures below 200°C. Anhydrous (dehydrated) ionic compounds have high melting points, usually above 300°C but below 1000°C. They are not readily volatilized and boil only at very high temperatures (Table 16.1).

When molten, ionic compounds conduct an electric current. In the solid state they do not conduct electricity. The conductivity in the molten liquid is attributed to the freedom of motion of the ions, which arises when the crystal lattice is no longer present.

Ionic substances are frequently but not always appreciably soluble in water. The solutions produced conduct the electric current rather well. The conductivity of a solution of a slightly soluble ionic substance is often several times that of the solvent water. Ionic substances are usually not nearly so soluble in other liquids as they are in water. For a liquid to be a good solvent for ionic compounds it must be highly polar, containing molecules with well-defined positive and negative regions with which the ions can interact.

MOLECULAR SUBSTANCES

All gases and essentially all liquids at room temperature are molecular in nature. If the molar mass of a substance is over about 100 grams, it may be a solid at room temperature. The

Table 16.1   Physical Properties of Some Representative Chemical Substances

Substance	M.P., °C	B.P., °C	Solubility		Electrical Conductance	Classification
			Water	Hexane		
NaCl	801	1413	Sol	Insol	High in melt and in soln	Ionic
MgO	2800	—	Sl sol	Insol	Low in sat'd soln	Ionic
$CoCl_2$	Sublimes	1049	Sol	Insol	High in soln	Ionic
$CoCl_2 \cdot 6H_2O$	86	Dec	Sol	Insol	High in soln	Ionic hydrate, $-H_2O$ at 110°C
$C_{10}H_8$	70	255	Insol	Sol	Zero in melt	Molecular
C_6H_5COOH	122	249	Sl sol	Sl sol	Low in sat'd soln	Molecular-ionic
$FeCl_3$	282	315	Sol	Sl sol	High in soln and in melt	Molecular-ionic
SnI_4	144	341	Dec	Sol	$\sim$ Zero in melt	Molecular
SiO_2	1600	2590	Insol	Insol	Zero in solid	Macromolecular
Fe	1535	3000	Insol	Insol	High in solid	Metallic

Key: Sol = at least 0.1 mole/L; Sl sol = appreciable solubility but < 0.1 mole/L; Insol = essentially insoluble; Dec = decomposes.

melting points of molecular substances are usually below 300°C; these substances are relatively volatile, but a good many will decompose before they boil. Most molecular substances do not conduct the electric current either when solid or when molten.

Organic compounds, which contain primarily carbon and hydrogen, often in combination with other nonmetals, are essentially molecular in nature. Since there are a great many organic substances, it is true that most substances are molecular. If an organic compound decomposes on heating, the residue is frequently a black carbonaceous material. Reasonably large numbers of inorganic substances are also molecular; those that are solids at room temperature include some of the binary compounds of elements in Groups 4, 5, 6, and 7.

Molecular substances are usually soluble in at least a few organic solvents, with the solubility being enhanced if the substance and the solvent are similar in molecular structure.

Some molecular compounds are markedly polar, which tends to increase their solubility in water and other polar solvents. Such substances may ionize appreciably in water, or even in the melt, so that they become conductors of electricity. Usually the conductivity is considerably lower than that of an ionic material. Most polar molecular compounds in this category are organic, but a few, including some of the salts of the transition metals, are inorganic.

MACROMOLECULAR SUBSTANCES

Macromolecular substances are all solids at room temperature. They have very high melting points, usually above 1000°C, and low volatility. They are typically very resistant to thermal decomposition. They do not conduct electric current and are often good insulators. They are not soluble in water or any organic solvents. They are frequently chemically inert and may be used as abrasives or refractories.

METALLIC SUBSTANCES

The properties of metals appear to derive mainly from the freedom of movement of their bonding electrons. Metals are good electrical conductors in the solid form, and have characteristic luster and malleability. Most metals are solid at room temperature and have melting points that range from below 0°C to over 2000°C. They are not soluble in water or organic solvents. Some metals are prepared as gray or black powders, which may not appear to be electrical conductors. However, if one measures the conductance while the powder is under pressure, its metallic character is revealed.

Experimental Procedure

In this experiment you will investigate the properties of several substances with the purpose of determining whether they are molecular, ionic, macromolecular, or metallic. In some cases the classification will be very straightforward. In others the assignment to a class will not be so easy and you may find that the substance has a behavior associated with more than one class.

As the discussion indicates, there are several properties we can use to find out to which class a substance belongs. In this experiment we will use the melting point, solubility in water and organic solvents, and electrical conductivity of the aqueous solution, the solid, and the melt in making the classification.

The substances to be studied in the first part of the experiment are on the laboratory tables along with two organic solvents, one polar and one nonpolar. You need only carry out enough tests on each substance to establish the class to which it belongs, so you will not need to perform every test on every substance. You may, however, carry out any extra tests you wish, if only to satisfy your curiosity. Follow the directions for each test as given below.

Melting Point

Approximate melting points of substances can be determined rather easily. Substances with low melting points, less than 100°C, will melt readily when warmed gently in a small test tube. A sample the size of a pea will suffice. If the sample melts between 100° and 300°C, it will take more than gentle warming, but will melt before the test tube imparts a yellow-orange color to the Bunsen flame. Above 300°C, there will be increasing color; up to about 500°C one can still use a test tube and a strong burner flame; but at about 550°C the pyrex tube will begin to soften. In this experiment we will not attempt to measure any melting points above 500°C.

While heating a sample, keep the tube *loosely* stoppered with a cork. *Do not breathe* any vapors that are given off, and *do not continue to heat* a sample after it has melted. As you heat the sample, look for evidence of decomposition, sublimation, or evolution of water.

Solubility and Conductance of Solutions

In testing for solubility, again use a sample about the size of a pea, this time in a regular test tube. Use about 2 mL of solvent, enough to fill the tube to a depth of about 1/2 inch. Stir well, using a clean stirring rod. Some samples will dissolve completely almost immediately; some are only slightly soluble and may produce a cloudy suspension; others are completely insoluble. Make solubility tests with distilled water and the two organic solvents and record your results. Use fresh distilled water in your wash bottle for the solubility tests.

Conductance measurements need only be made on water solutions. We will use portable ohm-meters for this purpose. Your instructor may make the measurements for you, but may have you make them. An ohm-meter measures the electrical resistance of a sample in ohms, Ω. A solution with a high resistance has a low electrical conductance, and vice versa. Some of your solutions will have a low resistance, of the order of 1000Ω or less; these are good conductors. Distilled water has a high resistance; with your meter it will probably have a resistance of 50,000Ω or greater. Small amounts of contaminants can lower the resistance of a solution very markedly, particularly if the main solute shows high resistance.

Measure the resistance of any of the aqueous solutions containing soluble or slightly soluble substances. Between tests, rinse the electrodes in a beaker filled with distilled water. For our purposes, a solution with a resistance less than about 2000Ω is a good conductor, G. Between 2000 and 20,000Ω it is a weak conductor, W. Above 20,000Ω we will consider it to be essentially nonconducting, N. Record the resistances you observe in ohms. Then note, with a G, W, or N, whether the solution is a good, weak, or poor conductor.

Electrical Conductance of Solids and Melts

Some substances conduct electricity in the solid state. If the sample contains large crystals, the conductance test is very easy. Select a crystal, put it on the lab bench, and touch it with the two wires on the ohm-meter probe. Metals have a very low resistance. In powder form most substances, metals included, appear to have essentially infinite resistance. However, under pressure, metal powders, unlike those of other substances, show good conductance. To test a powder for conductance, put a penny on the lab bench. On it place a small rubber washer from the box on the bench. Fill the hole in the washer with the powder, and put another penny on top of the washer. Put the whole sandwich between the jaws of a pair of insulated pliers. Touch the electrodes from the ohm-meter probe to the pennies, one electrode to each penny, and squeeze the pliers. If the powder is a metal, the resistance will gradually fall from infinity to a small value. Make sure any drop in resistance is not caused by the pennies touching each other. Record your results.

To check the conductance of a melt, put a pea-size sample in a dry, regular test tube and melt it. Heat the electrodes on the probe for a few seconds in the Bunsen flame and touch them to the melt. Heat gently to ensure that no solid is crystallized on the electrodes. Many melts are good conductors. After testing a melt, clean the electrodes by washing them with water or an organic solvent, or, if necessary, scraping them off with a spatula.

Having made the tests we have described, you should be able to assign each substance to its class, or, possibly, to one or both of two classes. Make this classification for each substance, and give your reasons for doing so.

When you have classified each substance, report to your laboratory supervisor, who will assign you two unknowns for characterization.

All residues from your tests should be discarded in the waste crock unless directed otherwise by your instructor.

Experiment 16

Observations and Conclusions: Classification of Chemical Substances

Table 16.2

Substance No.	Approx. Melting Pt., °C (<100, $100-300$, $300-500$, >500)	Solubility			Electrical resistance, Ω			Classification and Reason
		H_2O	Nonpolar Organic	Polar Organic	Solution in H_2O	Solid	Melt	
I								
II								
III								
IV								
V								
VI								
Unknown No. ___ ___								

Key: Sol = soluble; Sl sol = slightly soluble; Insol = insoluble; G = good electrical conductor, R < 2000Ω; W = weak electrical conductor, 2000Ω < R < 20,000Ω; N = nonconductor, R > 20,000Ω.

Advance Study Assignment: Classification of Substances

1. List the properties of a substance that would definitely establish that the material is molecular.

2. If we classify substances as ionic, molecular, macromolecular, or metallic, in which if any categories are all the members

    a. soluble in water?

    b. electrical conductors in the melt?

    c. insoluble in all common solvents?

    d. solids at room temperature?

3. A given substance is a white solid at 25°C. It melts at 350°C without decomposing, and the melt conducts an electric current. What would be the classification of the substance, based on this information?

4. A white solid melts at 1000°C. The melt does not conduct electricity. Classify the substance as best you can from these properties.

Experiment 17

MOLAR MASS DETERMINATION BY DEPRESSION OF THE FREEZING POINT

In an earlier experiment you observed the change of vapor pressure of a liquid as a function of temperature. If a nonvolatile solid compound (the solute) is dissolved in a liquid, the vapor pressure of the liquid solvent is lowered. This decrease in the vapor pressure of the solvent results in other easily observable physical changes; the boiling point of the solution is higher than that of the pure solvent, and the freezing point is lower.

Many years ago chemists observed that at low solute concentrations the changes in the boiling point, the freezing point, and the vapor pressure of a solution are all proportional to the amount of solute that is dissolved in the solvent. These three properties are collectively known as colligative properties of solutions. The colligative properties of a solution depend only on the number of solute particles present in a given amount of solvent and not on the kind of particles dissolved.

When working with boiling point elevations or freezing point depressions of solutions, it is convenient to express the solute concentration in terms of its molality m defined by the relation:

$$\text{molality of } A = m_A = \frac{\text{no. of moles } A \text{ dissolved}}{\text{no. of kg solvent in the solution}}$$

For this unit of concentration, the boiling point elevation, $T_b - T_b^\circ$ or ΔT_b, and the freezing point depression, $T_f^\circ - T_f$ or ΔT_f, in °C at low concentrations are given by the equations:

$$\Delta T_b = k_b m \qquad \Delta T_f = k_f m \tag{1}$$

where k_b and k_f are characteristic of the solvent used. For water, $k_b = 0.52$ and $k_f = 1.86$. For benzene, $k_b = 2.53$ and $k_f = 5.10$.

One of the main uses of the colligative properties of solutions is in connection with the determination of the molar masses of unknown substances. If we dissolve a known amount of solute in a given amount of solvent and measure ΔT_b or ΔT_f of the solution produced, and if we know the appropriate k for the solvent, we can find the molality and hence the molar mass, MM, of the solute. In the case of the freezing point depression, the relation would be:

$$\Delta T_f = k_f m = k_f \times \frac{\text{no. moles solute}}{\text{no. kg solvent}} = k_f \times \frac{\left(\dfrac{\text{no. g solute}}{\text{MM solute}}\right)}{\text{no. kg solvent}} \tag{2}$$

In this experiment you will be asked to estimate the molar mass of an unknown solute, using this equation. The solvent used will be stearic acid, which has a convenient melting point and a moderately large value for k_f, 4.89. The freezing points will be obtained by studying the rate at which liquid stearic acid and some of its solutions containing the unknown cool in air.

When a pure substance that melts at, say, 70°C is heated to 80°C, where it will be completely liquid, and then allowed to cool in air, the temperature of the sample will vary with time, as in Figure 17.1. Initially the temperature will fall quite rapidly. When the freezing

FIGURE 17.1

point is reached, solid will begin to form, and the temperature will tend to hold steady until the sample is all solid. The freezing point of the pure liquid is the constant temperature observed while the liquid is freezing to a solid.

The cooling behavior of a solution is somewhat different from that of a pure liquid, and is also shown in Figure 17.1. The temperature at which the solution begins to freeze is lower than for the pure solvent. In addition, there is a slow gradual fall in temperature as freezing proceeds. The best value for the freezing point of the solution is obtained by drawing two straight lines connecting the points on the temperature-time graph. The first line connects points where the solution is all liquid. The second line connects points where solid and liquid coexist. The point where the two lines intersect is the freezing point of the solution. With both the pure liquid and solutions, at the time when solid first appears, the temperature may fall below the freezing point and then come back up to it as solid forms. The effect is called supercooling, and is shown in Figure 17.1. When drawing the straight line in the solid-liquid region of the graph, ignore points where supercooling was observed. To establish the proper straight line in the solid-liquid region, it is necessary to record the temperature until the trend with time is smooth and clearly established.

Experimental Procedure

WEAR YOUR SAFETY GLASSES WHILE PERFORMING THIS EXPERIMENT

A. Determination of the Freezing Point of Stearic Acid

From the stockroom obtain a stopper fitted with a sensitive thermometer and a glass stirrer, a large test tube, and a sample of solid unknown. *Remember that the thermometer is both fragile and expensive, so handle it with due care.* Weigh the test tube on a top-loading or triple-beam balance to 0.01 g. Add about 20 g of stearic acid, SA, to the test tube and weigh again to the same precision.

Fill your 600-mL beaker 3/4 full of hot water from the faucet. Support the beaker on an iron ring and wire gauze on a ring stand and heat the water with a Bunsen burner. Clamp the test tube to the ring stand and immerse the tube in the water as far as is convenient (see Fig. 17.2).

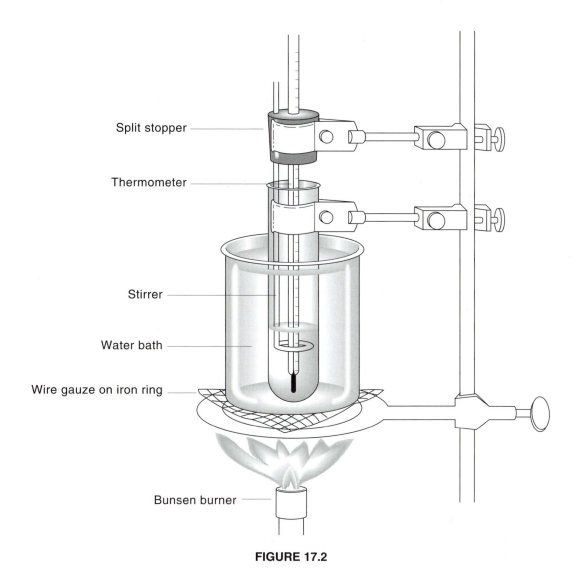

FIGURE 17.2

When the water gets to the melting point of the stearic acid, it will slowly begin to melt. The solid is quite fluffy, so it will be helpful if you gently push it down with your stirrer so it can make better contact with the tube wall. If the water bath reaches the boiling point, remove heat. Stir until the SA is all melted. Assemble the thermometer-stirrer assembly, and clamp it as shown in Figure 17.2, adjusting the level of the thermometer bulb so that it is about 1 cm above the bottom of the tube. Note the temperature of the melt, which should be about 80°C, or a little higher. If necessary, use your burner to bring it to that temperature and then remove heat. Let the beaker cool until the top edge is comfortable to hold. Then, carefully, using a paper towel or a handkerchief, loosen the hot ring clamp, lower the iron ring and water bath, and put the beaker of hot water on the lab bench, well away from the test tube.

Record the temperature of the stearic acid as it cools in the air. Stir the liquid slowly but continuously to minimize supercooling. Start readings at about 78°C and note the temperature every 30 seconds for 8 minutes or until the liquid has solidified to the point that you are no longer able to stir it. Near the melting point you will begin to observe crystals of SA in the liquid, and these will increase in amount as cooling proceeds.

B. Determination of the Molar Mass of an Unknown Compound

Weigh your unknown in its container to 0.01 g. Pour about half of the sample (about 2 g) into your test tube of SA and reweigh the container. Be careful with both weighings.

Heat the test tube in the water bath until the SA is again melted and the solid unknown is dissolved. Make sure that all the SA on the walls is melted; stir well to mix the unknown with the SA thoroughly. When the melt has reached about 80°C, remove the bath, and let it cool as before. Start readings at about 78°C and continue to take readings, with stirring, for 8 minutes.

The dependence of temperature on time with the solution will be similar to that observed for pure SA, except that the first crystals will appear at a lower temperature, and the temperature of the solid-solution system will gradually fall as cooling proceeds. There may be some supercooling, as evidenced by a rise in temperature shortly after the first apearance of SA crystals.

Add the rest of your unknown (about 2 g) to the SA solution and again weigh the container. Melt the SA as before, heating it to about 80°C before removing the water bath. Start temperature-time readings at about 75°C. Repeat the entire procedure described above.

When you have completed the experiment, melt the SA solution and pour it into the waste crock. The residue in the test tube can be cleaned out with soap and hot water.

Data and Calculations: Molar Mass Determination by Freezing Point Depression

Mass of large test tube   _____ g

Mass of test tube plus about 20 g of stearic acid   _____ g

Mass of container plus unknown   _____ g

Mass of container less Sample I   _____ g

Mass of container less Sample II   _____ g

Time-temperature readings

Time (minutes)	Temperature, °C		
	Stearic Acid	*Solution I*	*Solution II*
0	_____	_____	_____
1/2	_____	_____	_____
1	_____	_____	_____
1 1/2	_____	_____	_____
2	_____	_____	_____
2 1/2	_____	_____	_____
3	_____	_____	_____
3 1/2	_____	_____	_____
4	_____	_____	_____
4 1/2	_____	_____	_____
5	_____	_____	_____
5 1/2	_____	_____	_____

(continued on following page)

(continued)

6	_____	_____	_____
6 1/2	_____	_____	_____
7	_____	_____	_____
7 1/2	_____	_____	_____
8	_____	_____	_____

Estimation of Freezing Point

On the graph paper provided, plot your temperature vs. time readings for pure stearic acid and for each run of the two solutions. To avoid overlapping graphs, add 4 minutes to all observed times in making the graph of the cooling curve for pure stearic acid. Add 2 minutes to all times for Solution I. Use times as observed for Solution II. Connect the points on each cooling curve with two straight lines, ignoring points involving supercooling. The freezing point is at the point where the two lines intersect.

Freezing points (from graphs)

_____	_____	_____
Pure SA	Solution I	Solution II

Calculation of Molar Mass

	Solution I	*Solution II*
Mass of unknown used (*total* amount in solution)	_____ g	_____ g
Mass of stearic acid used	_____ g	_____ g
Freezing point of pure stearic acid	_____ °C	_____ °C
Freezing point of solution	_____ °C	_____ °C
Freezing point depression (total depression)	_____ °C	_____ °C
Total molal concentration of unknown solution (from Eq. 1; $k_f = 4.89$)	_____	_____
Molar mass of unknown (from Eq. 2)	_____ g	_____ g
Average molar mass		_____ g
Unknown no.		_____

(continued on following page)

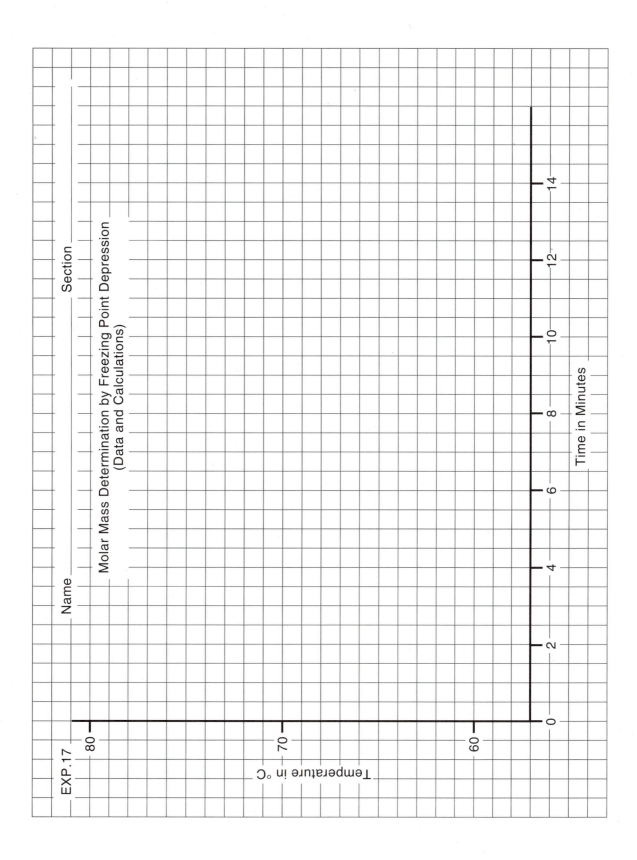

EXP. 17

Molar Mass Determination by Freezing Point Depression
(Data and Calculations)

Name

Section

Temperature in °C

Time in Minutes

Advance Study Assignment: Determination of Molar Mass by Freezing Point Depression

1. A student determines the freezing point of a solution of 2.07 g of naphthalene in 21.00 g of stearic acid. He obtains the following temperature-time readings:

Time (min)	0.0	0.5	1.0	1.5	2.0	2.5	3.0	3.5	4.0
Temp (°C)	79.7	77.4	75.2	73.0	70.4	66.6	66.3	64.4	64.3

Time (min)	4.5	5.0	5.5	6.0	6.5	7.0	7.5	8.0
Temp (°C)	64.3	64.2	64.2	63.9	63.8	63.7	63.6	63.5

a. Plot these data on the graph paper provided. Note that the first several points lie essentially on a straight line. Similarly, the last several points lie on another straight line. Draw in those two lines. The point at which those lines intersect is the freezing point of the solution.

b. What is the freezing point of the solution?

————————— °C

c. What is the freezing point depression, $T_f^\circ - T_f$, or ΔT_f? Take T_f° to be 68.2°C for stearic acid.

————————— °C

d. What is the molality of the naphthalene? (Use Eq. 1: $k_f = 4.89$)

————————— m

e. What is the molar mass of naphthalene? Use Equation 2, as modified below:

$$MM = \frac{\text{no. g solute}}{\text{no. kg solvent}} \times \frac{k_f}{\Delta T_f}$$ ————————— g

f. The molecular formula of naphthalene is $C_{10}H_8$. Is the result of Part e consistent with this formula?

—————————

(continued on following page)

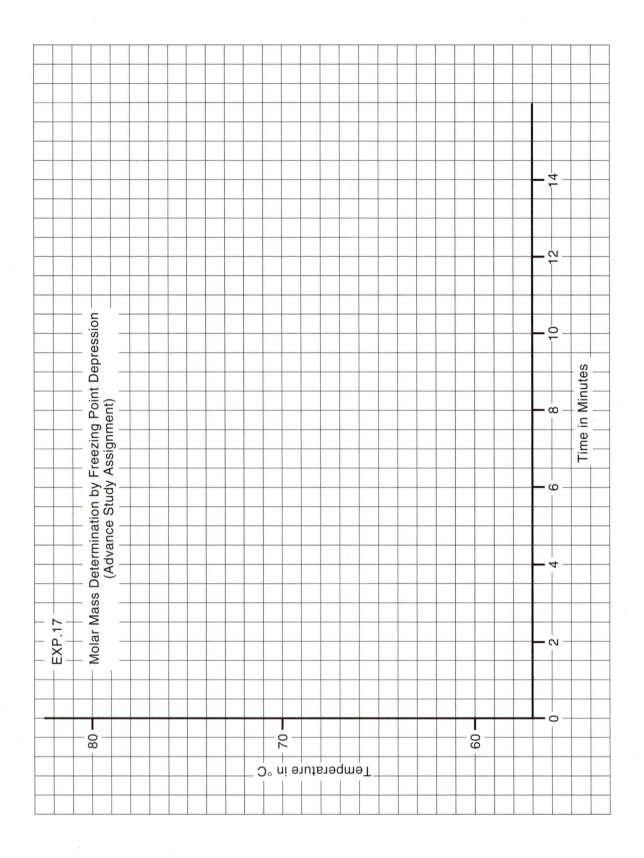

EXP.17

Molar Mass Determination by Freezing Point Depression
(Advance Study Assignment)

Temperature in °C

80

70

60

Time in Minutes

0   2   4   6   8   10   12   14

Experiment 18

RATES OF CHEMICAL REACTIONS, I. THE IODINATION OF ACETONE

The rate at which a chemical reaction occurs depends on several factors: the nature of the reaction, the concentrations of the reactants, the temperature, and the presence of possible catalysts. All of these factors can markedly influence the observed rate of reaction.

Some reactions at a given temperature are very slow indeed; the oxidation of gaseous hydrogen or wood at room temperature would not proceed appreciably in a century. Other reactions are essentially instantaneous; the precipitation of silver chloride when solutions containing silver ions and chloride ions are mixed and the formation of water when acidic and basic solutions are mixed are examples of extremely rapid reactions. In this experiment we will study a reaction that, in the vicinity of room temperature, proceeds at a moderate, relatively easily measured rate.

For a given reaction, the rate typically increases with an increase in the concentration of any reactant. The relation between rate and concentration is a remarkably simple one in many cases, and for the reaction

$$aA + bB \rightarrow cC$$

the rate can usually be expressed by the equation

$$\text{rate} = k(A)^m(B)^n \tag{1}$$

where m and n are generally, but not always, integers, 0, 1, 2, or possibly 3; (A) and (B) are the concentrations of A and B (ordinarily in moles per liter); and k is a constant, called the *rate constant* of the reaction, which makes the relation quantitatively correct. The numbers m and n are called the *orders of the reaction* with respect to A and B. If m is 1, the reaction is said to be *first order* with respect to the reactant A. If n is 2, the reaction is *second order* with respect to reactant B. The overall order is the sum of m and n. In this example the reaction would be *third order* overall.

The rate of a reaction is also significantly dependent on the temperature at which the reaction occurs. An increase in temperature increases the rate, an often-cited rule being that a 10°C rise in temperature will double the rate. This rule is only approximately correct; nevertheless, it is clear that a rise of temperature of say 100°C could change the rate of a reaction appreciably.

As with the concentration, there is a quantitative relation between reaction rate and temperature, but here the relation is somewhat more complicated. This relation is based on the idea that to react, the reactant species must have a certain minimum amount of energy present at the time the reactants collide in the reaction step; this amount of energy, which is typically furnished by the kinetic energy of motion of the species present, is called the *activation energy* for the reaction. The equation relating the rate constant k to the absolute temperature T and the activation energy E_a is

$$\log_{10} k = \frac{-E_a}{2.30RT} + \text{constant} \tag{2}$$

where R is the gas constant (8.31 joules/mole K for E_a in joules per mole). This equation is identical in form to Equation 1 in Experiment 14. By measuring k at different temperatures we can determine graphically the activation energy for a reaction.

In this experiment we will study the kinetics of the reaction between iodine and acetone:

$$CH_3-\overset{\displaystyle O}{\overset{\displaystyle \|}{C}}-CH_3(aq) + I_2(aq) \rightarrow CH_3-\overset{\displaystyle O}{\overset{\displaystyle \|}{C}}-CH_2I(aq) + H^+(aq) + I^-(aq)$$

The rate of this reaction is found to depend on the concentration of hydrogen ion in the solution as well as presumably on the concentrations of the two reactants. By Equation 1, the rate law for this reaction is

$$\text{rate} = k(\text{acetone})^m(I_2)^n(H^+)^p \tag{3}$$

where m, n, and p are the orders of the reaction with respect to acetone, iodine, and hydrogen ion, respectively, and k is the rate constant for the reaction.

The rate of this reaction can be expressed as the (small) change in the concentration of I_2, $\Delta(I_2)$, that occurs, divided by the time interval Δt required for the change:

$$\text{rate} = \frac{-\Delta(I_2)}{\Delta t} \tag{4}$$

The minus sign is to make the rate positive ($\Delta(I_2)$ is negative). Ordinarily, since rate varies as the concentrations of the reactants according to Equation 3, in a rate study it would be necessary to measure, directly or indirectly, the concentration of each reactant as a function of time; the rate would typically vary markedly with time, decreasing to very low values as the concentration of at least one reactant becomes very low. This makes reaction rate studies relatively difficult to carry out and introduces mathematical complexities that are difficult for beginning students to understand.

The iodination of acetone is a rather atypical reaction, in that it can be easily investigated experimentally. First of all, iodine has color, so that one can readily follow changes in iodine concentration visually. A second and very important characteristic of this reaction is that it turns out to be zero order in I_2 concentration. This means (see Equation 3) that the rate of the reaction does not depend on (I_2) at all; $(I_2)^0 = 1$, no matter what the value of (I_2) is, as long as it is not itself zero.

Because the rate of the reaction does not depend on (I_2), we can study the rate by simply making I_2 the limiting reagent present in a large excess of acetone and H^+ ion. We then measure the time required for a known initial concentration of I_2 to be used up completely. If both acetone and H^+ are present at much higher concentrations than that of I_2, their concentrations will not change appreciably during the course of the reaction, and the rate will remain, by Equation 3, effectively constant until all the iodine is gone, at which time the reaction will stop. Under such circumstances, if it takes t seconds for the color of a solution having an initial concentration of I_2 equal to $(I_2)_0$ to disappear, the rate of the reaction, by Equation 4, would be

$$\text{rate} = \frac{-\Delta(I_2)}{\Delta t} = \frac{(I_2)_0}{t} \tag{5}$$

Although the rate of the reaction is constant during its course under the conditions we have set up, we can vary it by changing the initial concentrations of acetone and H^+ ion. If, for example, we should *double* the initial concentration of *acetone* over that in Mixture 1, keeping (H^+) and (I_2) at the *same* values they had previously, then the rate of Mixture 2 would, according to Equation 3, be different from that in Mixture 1:

$$\text{rate } 2 = k(2A)^m(I_2)^0(H^+)^p \tag{6a}$$

$$\text{rate } 1 = k(A)^m(I_2)^0(H^+)^p \tag{6b}$$

Dividing the first equation by the second, we see that the k's cancel, as do the terms in the iodine and hydrogen ion concentrations, since they have the same values in both reactions, and we obtain simply

$$\frac{\text{rate } 2}{\text{rate } 1} = \frac{(2A)^m}{(A)^m} = \left(\frac{2A}{A}\right)^m = 2^m \tag{6}$$

Having measured both rate 2 and rate 1 by Equation 5, we can find their ratio, which must be equal to 2^m. We can then solve for m either by inspection or using logarithms and so find the *order* of the reaction with respect to acetone.

By a similar procedure we can measure the order of the reaction with respect to H^+ ion concentration and also confirm the fact that the reaction is zero order with respect to I_2. Having found the order with respect to each reactant, we can then evaluate k, the rate constant for the reaction.

The determination of the orders m and p, the confirmation of the fact that n, the order with respect to I_2, equals zero, and the evaluation of the rate constant k for the reaction at room temperature comprise your assignment in this experiment. You will be furnished with standard solutions of acetone, iodine, and hydrogen ion, and with the composition of one solution that will give a reasonable rate. The rest of the planning and the execution of the experiment will be your responsibility.

An optional part of the experiment is to study the rate of this reaction at different temperatures to find its activation energy. The general procedure here would be to study the rate of reaction in one of the mixtures at room temperature and at two other temperatures, one above and one below room temperature. Knowing the rates, and hence the k's, at the three temperatures, you can then find E_a, the energy of activation for the reaction, by plotting log k vs. $1/T$. The slope of the resultant straight line, by Equation 2, must be $-E_a/2.30R$.

Experimental Procedure

Select two regular test tubes; when filled with distilled water, they should appear to have identical color when you view them down the tubes against a white background.

Draw 50 mL of each of the following solutions into clean, dry, 100-mL beakers, one solution to a beaker: 4 M acetone, 1 M HCl, and 0.005 M I_2. Cover each beaker with a watch glass.

With your graduated cylinder, measure out 10.0 mL of the 4 M acetone solution and pour it into a clean 125-mL Erlenmeyer flask. Then measure out 10.0 mL 1 M HCl and add that to the acetone in the flask. Add 20.0 mL distilled H_2O to the flask. Drain the graduated cylinder, shaking out any excess water, and then use the cylinder to measure out 10.0 mL 0.005 M I_2 solution. Be careful not to spill the iodine solution on your hands or clothes.

Noting the time on your wristwatch or the wall clock to 1 second, pour the iodine solution into the Erlenmeyer flask and quickly swirl the flask to mix the reagents thoroughly. The reaction mixture will appear yellow because of the presence of the iodine, and the color will fade slowly as the iodine reacts with the acetone. Fill one of the test tubes 3/4 full with the reaction mixture, and fill the other test tube to the same depth with distilled water. Look down the test tubes toward a well-lit piece of white paper, and note the time the color of the iodine just disappears. Measure the temperature of the mixture in the test tube.

Repeat the experiment, using as a reference the reacted solution instead of distilled water. The amount of time required in the two runs should agree within about 20 seconds.

The rate of the reaction equals the initial concentration of I_2 *in the reaction mixture* divided by the elapsed time. Since the reaction is zero order in I_2, and since both acetone and H^+ ion are present in great excess, the rate is constant throughout the reaction and the concentrations of both acetone and H^+ remain essentially at their initial values in the reaction mixture.

Having found the reaction rate for one composition of the system, it might be well to think for a moment about what changes in composition you might make to decrease the time and hence increase the rate of reaction. In particular, how could you change the composition in such a way as to allow you to determine how the rate depends upon acetone concentration? If it is not clear how to proceed, reread the discussion preceding Equation 6. In your new mixture you should keep the total volume at 50 mL, and be sure that the concentrations of H^+ and I_2 are the *same* as in the first experiment. Carry out the reaction twice with your new mixture; the times should not differ by more than about 15 seconds. The temperature should be kept within about a degree of that in the initial run. Calculate the rate of the reaction. Compare it with that for the first mixture, and then calculate the order of the reaction with respect to acetone, using a relation similar to Equation 6. First, write an equation like 6a for the second reaction mixture, substituting in the values for the rate as obtained by Equation 5 and the initial concentration of acetone, I_2, and H^+ in the reaction mixture. Then write an equation like 6b for the first reaction mixture, using the observed rate and the initial concentrations in that mixture. Obtain an equation like 6 by dividing Equation 6a by Equation 6b. Solve Equation 6 for the order m of the reaction with respect to acetone.

Again change the composition of the reaction mixture so that this time a measurement of the reaction will give you information about the order of the reaction with respect to H^+. Repeat the experiment with this mixture to establish the time of reaction to within 15 seconds, again making sure that the temperature is within about a degree of that observed previously. From the rate you determine for this mixture find p, the order of the reaction with respect to H^+.

Finally, change the reaction mixture composition in such a way as to allow you to show that the order of the reaction with respect to I_2 is zero. Measure the rate of the reaction twice, and calculate n, the order with respect to I_2.

Having found the order of the reaction for each species on which the rate depends, evaluate k, the rate constant for the reaction, from the rate and concentration data in each of the mixtures you studied. If the temperatures at which the reactions were run are all equal to within a degree or two, k should be about the same for each mixture.

(OPTIONAL). As a final reaction, make up a mixture using reactant volumes that you did not use in any previous experiments. Using Equation 3, the values of concentrations in the mixtures, the orders, and the rate constant you calculated from your experimental data, predict how long it will take for the I_2 color to disappear from your mixture. Measure the time for the reaction and compare it with your prediction.

If time permits, select one of the reaction mixtures you have already used that gave a convenient time, and use that mixture to measure the rate of reaction at about 10°C and at about 40°C. From the two rates you find, plus the rate at room temperature, calculate the energy of activation for the reaction, using Equation 2.

The reagents used in this experiment are nonpolluting and may be discarded down the sink drain.

Data and Calculations: The Iodination of Acetone

A. Reaction Rate Data

Mixture	Volume in mL 4.0 M acetone	Volume in mL 1.0 M HCl	Volume in mL 0.0050 M I_2	Volume in mL H_2O	Time for reaction in sec 1st run	2nd run	Temp. °C
I	10	10	10	20	_____	_____	_____
II	_____	_____	_____	_____	_____	_____	_____
III	_____	_____	_____	_____	_____	_____	_____
IV	_____	_____	_____	_____	_____	_____	_____

B. Determination of Reaction Orders with Respect to Acetone, H⁺ Ion, and I_2

$$\text{rate} = k(\text{acetone})^m (I_2)^n (H^+)^p \tag{3}$$

Calculate the *initial* concentrations of acetone, H^+ ion, and I_2 in each of the mixtures you studied. Use Equation 5 to find the rate of each reaction.

Mixture	(acetone)	(H⁺)	$(I_2)_0$	rate $= \dfrac{(I_2)_0}{\text{avg. time}}$
I	0.80 M	0.20 M	0.0010 M	_____
II	_____	_____	_____	_____
III	_____	_____	_____	_____
IV	_____	_____	_____	_____

Substituting the initial concentrations and the rate from the table above, write Equation 3 as it would apply to Reaction Mixture II:

Rate II =

Now write Equation 3 for Reaction Mixture I, substituting concentrations and the calculated rate from the table:

Rate I =

Divide the equation for Mixture II by the equation for Mixture I; the resulting equation should have the ratio of Rate II to Rate I on the left side, and a ratio of acetone concentrations

(continued on following page)

(continued)

raised to the power m on the right. It should be similar in appearance to Equation 6. Put the resulting equation below:

$$\frac{\text{Rate II}}{\text{Rate I}} =$$

The only unknown in the equation is m. Solve for m. $m =$ _____

Now write Equation 3 as it would apply to Reaction Mixture III and as it would apply to Reaction Mixture IV:

Rate III =

Rate IV =

Using the ratios of the rates of Mixtures III and IV to those of Mixtures II or I, find the orders of the reaction with respect to H^+ ion and I_2:

$$\frac{\text{Rate III}}{\text{Rate} \underline{\hspace{2cm}}} = \qquad\qquad p = \underline{\hspace{2cm}}$$

$$\frac{\text{Rate IV}}{\text{Rate} \underline{\hspace{2cm}}} = \qquad\qquad n = \underline{\hspace{2cm}}$$

C. Determination of the Rate Constant k

Given the values of m, p, and n as determined in Part B, calculate the rate constant k for each mixture by simply substituting those orders, the initial concentrations, and the observed rate from the table into Equation 3.

Mixture	*I*	*II*	*III*	*IV*	*average*
k	_____	_____	_____	_____	_____

D. Prediction of Reaction Rate (Optional)

Reaction mixture

Volume in mL 4.0 M acetone _____	Volume in mL 1.0 M HCl _____	Volume in mL 0.0050 M I_2 _____	Volume in mL H_2O _____

Initial concentrations

(acetone) _____ M $(H+)$ _____ M $(I_2)_0$ _____ M

Predicted rate _____ (Eq. 3)

Predicted time for reaction _____ sec (Eq. 5)

Observed time for reaction _____ sec

(continued on following page)

E. Determination of Energy of Activation (Optional)

Reaction mixture used _____ (same for all temperatures)

Time for reaction at about 10°C _____ sec   temperature _____°C; _____ K

Time for reaction at about 40°C _____ sec   temperature _____°C; _____ K

Time for reaction at room temp _____ sec   temperature _____°C; _____ K

Calculate the rate constant at each temperature from your data, following the procedure in Part C.

	rate	k	*log* k	$\frac{1}{T(K)}$
~ 10°C	_____	_____	_____	_____
~ 40°C	_____	_____	_____	_____
room temp	_____	_____	_____	_____

Plot log k vs. $1/T$. Find the slope of the best straight line through the points. (If you need to, see Appendix V).

Slope = _____

By Equation 2:

$$E_a = -2.30 \times 8.31 \times \text{slope}$$

E_a = _____ joules

THE IODINATION OF ACETONE
(Data and Calculations)

Advance Study Assignment: The Iodination of Acetone

1. In a reaction involving the iodination of acetone, the following volumes were used to make up the reaction mixture:

   10 mL 4.0 M acetone + 10 mL 1.0 M HCl + 10 mL 0.0050 M I_2 + 20 mL H_2O

   a. How many moles of acetone were in the reaction mixture? Recall that, for a component A, no. moles $A = M_A \times V$, where M_A is the molarity of A and V the volume in liters of the solution of A that was used.

   _____ moles acetone

   b. What was the molarity of acetone in the *reaction mixture?* The volume of the *mixture* was 50 mL, 0.050 L, and the number of moles of acetone was found in Part a. Again,

   $$M_A = \frac{\text{no. moles } A}{V \text{ of soln. in liters}}$$

   _____ M acetone

   c. How could you double the molarity of the acetone in the reaction mixture, keeping the total volume at 50 mL and keeping the same concentrations of H^+ ion and I_2 as in the original mixture?

2. Using the reaction mixture in Problem 1, a student found that it took 250 seconds for the color of the I_2 to disappear.

   a. What was the rate of the reaction? Hint: First find the initial concentration of I_2 in the reaction mixture, $(I_2)_0$. Then use Equation 5.

   rate = _____

   b. Given the rate from Part a, and the initial concentrations of acetone, H^+ ion, and I_2 in the reaction mixture, write Equation 3 as it would apply to the mixture.

   rate =

   c. What are the unknowns that remain in the equation in Part b?

   _____ _____ _____ _____

(continued on following page)

(continued)

3. A second reaction mixture was made up in the following way:

    20 mL 4.0 M acetone + 10 mL 1.0 M HCl + 10 mL 0.0050 M I_2 + 10 mL H_2O

    a. What were the initial concentrations of acetone, H^+ ion, and I_2 in the reaction mixture?

    (acetone) _____ M; (H^+) _____ M; $(I_2)_0$ _____ M

    b. It took 120 seconds for the I_2 color to disappear from the reaction mixture when it occurred at the same temperature as the reaction in Problem 2. What was the rate of the reaction?

    _____

    Write Equation 3 as it would apply to the second reaction mixture:

    rate =

    c. Divide the equation in Part b by the equation in Problem 2b. The resulting equation should have the ratio of the two rates on the left side and a ratio of acetone concentrations raised to the m power on the right. Write the resulting equation and solve for the value of m, the order of the reaction with respect to acetone. (Round off the value of m to the nearest integer.)

    $m =$ _____

4. A third reaction mixture was made up in the following way:

    20 mL 4.0 M acetone + 10 mL 1.0 M HCl + 20 mL 0.0050 M I_2

    If the reaction is zero order in I_2, how long would it take for the I_2 color to disappear at the temperature of the reaction mixture in Problem 3?

    _____ seconds

Experiment 19

RATES OF CHEMICAL REACTIONS, II. A CLOCK REACTION

In the previous experiment we discussed the factors that influence the rate of a chemical reaction and presented the terminology used in quantitative relations in studies of the kinetics of chemical reactions. That material is also pertinent to this experiment and should be studied before you proceed further.

This experiment involves the study of the rate properties, or chemical kinetics, of the following reaction between iodide ion and bromate ion under acidic conditions:

$$6 \, I^-(aq) + BrO_3^-(aq) + 6 \, H^+(aq) \rightarrow 3 \, I_2(aq) + Br^-(aq) + 3 \, H_2O \qquad (1)$$

This reaction proceeds reasonably slowly at room temperature, its rate depending on the concentrations of the I^-, BrO_3^-, and H^+ ions according to the rate law discussed in the previous experiment. For this reaction the rate law takes the form

$$\text{rate} = k(I^-)^m(BrO_3^-)^n(H^+)^p \qquad (2)$$

One of the main purposes of the experiment will be to evaluate the rate constant k and the reaction orders m, n, and p for this reaction. We will also investigate the manner in which the reaction rate depends on temperature and will evaluate the activation energy E_a for the reaction.

Our method for measuring the rate of the reaction involves what is frequently called a clock reaction. In addition to Reaction 1, whose kinetics we will study, the following reaction will also be made to occur simultaneously in the reaction flask:

$$I_2(aq) + 2 \, S_2O_3^{2-}(aq) \rightarrow 2 \, I^-(aq) + S_4O_6^{2-}(aq) \qquad (3)$$

As compared with Equation 1 this reaction is essentially instantaneous. The I_2 produced in (1) reacts completely with the thiosulfate, $S_2O_3^{2-}$, ion present in the solution, so that until all the thiosulfate ion has reacted, the concentration of I_2 is effectively zero. As soon as the $S_2O_3^{2-}$ is gone from the system, the I_2 produced by (1) remains in the solution and its concentration begins to increase. The presence of I_2 is made strikingly apparent by a starch indicator that is added to the reaction mixture, since I_2 even in small concentrations reacts with starch solution to produce a blue color.

By carrying out Reaction 1 in the presence of $S_2O_3^{2-}$ and a starch indicator, we introduce a "clock" into the system. Our clock tells us when a given amount of BrO_3^- ion has reacted (1/6 mole BrO_3^- per mole $S_2O_3^{2-}$), which is just what we need to know, since the rate of reaction can be expressed in terms of the time it takes for a particular amount of BrO_3^- to be used up. In all our reactions, the amount of BrO_3^- that reacts in the time we measure will be constant and small as compared to the amounts of any of the other reactants. This means that the concentrations of all reactants will be essentially constant in Equation 2, and hence so will the rate during each reaction.

In our experiment we will carry out the reaction between BrO_3^-, I^-, and H^+ ions under different concentration conditions. Measured amounts of each of these ions in water solution will be mixed in the presence of a constant small amount of $S_2O_3^{2-}$. The time it takes for each mixture to turn blue will be measured. The time obtained for each reaction will be inversely proportional to its rate. By changing the concentration of one reactant and keeping the other concentrations constant, we can investigate how the rate of the reaction varies with the concentration of a particular reactant. Once we know the order for each reactant we can determine the rate constant for the reaction.

In the last part of the experiment we will investigate how the rate of the reaction depends on temperature. You will recall that in general the rate increases sharply with temperature. By measuring how the rate varies with temperature we can determine the activation energy, E_a, for the reaction by making use of the Arrhenius equation:

$$\log_{10}k = \frac{-E_a}{2.30RT} + \text{constant} \tag{4}$$

In this equation, k is the rate constant at the Kelvin temperature T, E_a is the activation energy, and R is the gas constant. By plotting $\log_{10}k$ against $1/T$ we should obtain, by Equation 4, a straight line whose slope equals $-E_a/2.30R$. From the slope of that line we can easily calculate the activation energy.

Experimental Procedure

WEAR YOUR SAFETY GLASSES WHILE
PERFORMING THIS EXPERIMENT

A. Dependence of Reaction Rate on Concentration

In Table 19.1 we have summarized the reagent volumes to be used in carrying out the several reactions whose rates we need to know to find the general rate law for Reaction 1. First, measure out 100 mL of each of the listed reagents (except H_2O) into clean, labeled flasks or beakers. Use these reagents in your reaction mixtures.

Table 19.1   Reaction Mixtures at Room Temperature
(Reagent Volumes in mL)

Reaction Mixture	Reaction Flask I (250 mL)			Reaction Flask II (125 mL)	
	0.010 M KI	0.0010 M Na₂S₂O₃	H₂O	0.040 M KBrO₃	0.10 M HCl
1	10	10	10	10	10
2	20	10	0	10	10
3	10	10	0	20	10
4	10	10	0	10	20
5	8	10	12	5	15

The actual procedure for each reaction mixture will be much the same, and we will describe it now for Reaction Mixture 1.

Since there are several reagents to mix, and since we don't want the reaction to start until we are ready, we will put some of the reagents into one flask and the rest into another, selecting them so that no reaction occurs until the contents of the two flasks are mixed. Using a 10-mL graduated cylinder to measure volumes, measure out 10 mL 0.010 M KI, 10 mL 0.0010 M Na₂S₂O₃, and 10 mL distilled water into a 250-mL Erlenmeyer flask (Reaction Flask I). Then measure out 10 mL 0.040 M KBrO₃ and 10 mL 0.10 M HCl into a 125-mL Erlenmeyer flask (Reaction Flask II). To Flask II add three or four drops of starch indicator solution.

Pour the contents of Reaction Flask II into Reaction Flask I and swirl the solutions to mix them thoroughly. Note the time at which the solutions were mixed. Continue swirling the solution. It should turn blue in less than 2 minutes. Note the time at the instant that the blue color appears. Record the temperature of the blue solution to 0.2°C.

Repeat the procedure with the other mixtures in Table 19.1. **Don't forget to add the indicator** before mixing the solutions in the two flasks. The reaction flasks should be rinsed with distilled water between runs. When measuring out reagents, rinse the graduated cylinder with distilled water after you have added the reagents to Reaction Flask I, and before you measure out the reagents for Reaction Flask II. Try to keep the temperature just about the same in all the runs. Repeat any experiments that did not appear to proceed properly.

B. Dependence of Reaction Rate on Temperature

In this part of the experiment, the reaction will be carried out at several different temperatures, using Reaction Mixture 1 in all cases. The temperatures we will use will be about 20°C, 40°C, 10°C, and 0°C.

We will take the time at about 20°C to be that for Reaction Mixture 1 as determined at room temperature. To determine the time at 40°C proceed as follows. Make up Reaction Mixture 1 as you did in Part A, including the indicator. However, instead of mixing the solutions in the two flasks at room temperature, put the flasks into water at 40°C, drawn from the hot-water tap into one or more large beakers. Check to see that the water is indeed at about 40°C, and leave the flasks in the water for several minutes to bring them to the proper temperature. Then mix the two solutions, noting the time of mixing. Continue swirling the reaction flask in the warm water. When the color change occurs, note the time and the temperature of the solution in the flask.

Repeat the experiment at about 10°C, cooling all the reactants in water at that temperature before starting the reaction. Record the time required for the color to change and the final temperature of the reaction mixture. Repeat once again at about 0°C, this time using an ice-water bath to cool the reactants.

C. Dependence of the Reaction Rate on the Presence of Catalyst (Optional)

Some ions have a pronounced catalytic effect on the rates of many reactions in water solution. Observe the effect on this reaction by once again making up Reaction Mixture 1. Before mixing, add one drop 0.5 M $(NH_4)_2MoO_4$, ammonium molybdate, and a few drops of starch indicator to Reaction Flask II. Swirl the flask to mix the catalyst thoroughly. Then mix the solutions, noting the time required for the color to change.

The reaction products in this experiment are very dilute and may be poured into the sink as you complete each part of the experiment.

Data and Calculations: Rates of Chemical Reactions, II. A Clock Reaction

A. Orders of the Reaction. Rate Constant Determination

Reaction: $6 I^-(aq) + BrO_3^-(aq) + 6 H^+(aq) \rightarrow 3 I_2(aq) + Br^-(aq) + 3 H_2O$ (1)

$$\text{rate} = k(I^-)^m(BrO_3^-)^n(H^+)^p = -\frac{\Delta(BrO_3^-)}{t} \qquad (2)$$

In all the reaction mixtures used in the experiment, the color change occurred when a constant predetermined number of moles of BrO_3^- had been used up by the reaction. The color "clock" allows you to measure the *time required* for this *fixed number of moles of* BrO_3^- *to react*. The rate of each reaction is determined by the time t required for the color to change; since in Equation 2 the change in concentration of BrO_3^- ion, $\Delta(BrO_3^-)$, is the same in each mixture, the relative rate of each reaction is inversely proportional to the time t. Since we are mainly concerned with relative rather than absolute rate, we will for convenience take all relative rates as being equal to $1000/t$. Fill in the following table, first calculating the relative reaction rate for each mixture.

Reaction Mixture	Time t (sec) for Color to Change	Relative Rate of Reaction 1000/t	Reactant Concentrations in Reacting Mixture (M)			Temp. in °C
			(I⁻)	(BrO₃⁻)	(H⁺)	
1	_____	_____	0.0020	_____	_____	_____
2	_____	_____	_____	_____	_____	_____
3	_____	_____	_____	_____	_____	_____
4	_____	_____	_____	_____	_____	_____
5	_____	_____	_____	_____	_____	_____

The reactant concentrations in the reaction mixture are *not* those of the stock solutions, since the reagents were diluted by the other solutions. The final volume of the reaction mixture is 50 mL in all cases. Since the number of moles of reactant does not change on dilution we can say, for example, for I^- ion, that

$$\text{no. moles } I^- = (I^-)_{stock} \times V_{stock} = (I^-)_{mixture} \times V_{mixture}$$

For Reaction Mixture 1,

$$(I^-)_{stock} = 0.010 \text{ M}, \ V_{stock} = 10 \text{ mL}, \ V_{mixture} = 50 \text{ mL}$$

Therefore,

$$(I^-)_{mixture} = \frac{0.010 \text{ M} \times 10 \text{ mL}}{50 \text{ mL}} = 0.0020 \text{ M}$$

Calculate the rest of the concentrations in the table by the same approach.

(continued on following page)

(continued)

Determination of the Orders of the Reaction

Given the data in the table, the problem is to find the order for each reactant and the rate constant for the reaction. Since we are dealing with relative rates, we can modify Equation 2 to read as follows:

$$\text{relative rate} = k'(\text{I}^-)^m(\text{BrO}_3^-)^n(\text{H}^+)^p \tag{5}$$

We need to determine the relative rate constant k' and the orders m, n, and p in such a way as to be consistent with the data in the table.

The solution to this problem is quite simple, once you make a few observations on the reaction mixtures. Each mixture (2 to 4) differs from Reaction Mixture 1 in the concentration of only one species (see table). This means that for any pair of mixtures that includes Reaction Mixture 1, there is only one concentration that changes. From the ratio of the relative rates for such a pair of mixtures we can find the order for the reactant whose concentration was changed. Proceed as follows.

Write Equation 5 below for Reaction Mixtures 1 and 2, substituting the relative rates and the concentrations of I^-, BrO_3^-, and H^+ ions from the table you have just completed.

Relative rate 1 = _____ = $k'($ $)^m($ $)^n($ $)^p$

Relative rate 2 = _____ = $k'($ $)^m($ $)^n($ $)^p$

Divide the first equation by the second, noting that nearly all the terms cancel out. The result is simply

$$\frac{\text{Relative rate 1}}{\text{Relative rate 2}} =$$

If you have done this properly, you will have an equation involving only m as an unknown. Solve this equation for m, the order of the reaction with respect to I^- ion.

$$m = \underline{\hspace{2cm}} \text{ (nearest integer)}$$

Applying the same approach to Reaction Mixtures 1 and 3, find the value of n, the order of the reaction with respect to BrO_3^- ion.

Relative rate 1 = _____ = $k'($ $)^m($ $)^n($ $)^p$

Relative rate 3 = _____ = $k'($ $)^m($ $)^n($ $)^p$

Dividing one equation by the other:

$$=$$

$$n = \underline{\hspace{2cm}}$$

Now that you have the idea, apply the method once again, this time to Reaction Mixtures 1 and 4, and find p, the order with respect to H^+ ion.

Relative rate 4 = $k'($ $)^m($ $)^n($ $)^p$

Dividing the equation for Relative Rate 1 by that for Relative Rate 4, we get

$$=$$

$$p = \underline{\hspace{2cm}}$$

(continued on following page)

(continued)

Having found m, n, and p (nearest integers), the relative rate constant, k', can be calculated by substitution of m, n, p, and the known rates and reactant concentrations into Equation 5. Evaluate k' for Reaction Mixtures 1 to 4.

Reaction	1	2	3	4	
k'	————	————	————	————	k'_{ave} ————

Why should k' have nearly the same value for each of the above reactions?

Using k'_{ave} in Equation 5, predict the relative rate and time t for Reaction Mixture 5. Use the concentrations in the table.

Relative rate$_{pred}$ ———— t_{pred} ———— t_{obs} ————

B. Effect of Temperature on Reaction Rate: The Activation Energy

To find the activation energy for the reaction it will be helpful to complete the following table. The dependence of the rate constant, k', for a reaction is given by Equation 4:

$$\log_{10}k' = \frac{-E_a}{2.3RT} + \text{constant} \tag{4}$$

Since the reactions at the different temperatures all involve the same reactant concentrations, the rate constants, k', for two different mixtures will have the same ratio as the reaction rates themselves for the two mixtures. This means that in the calculation of E_a, we can use the observed relative rates instead of rate constants. Proceeding as before, calculate the relative rates of reaction in each of the mixtures and enter these values in (c). Take the $\log_{10}$ rate for each mixture and enter these values in (d). To set up the terms in $1/T$, fill in (b), (e), and (f) in the table.

	Approximate temperature in °C			
	20	40	10	0
(a) Time t in seconds for color to appear	————	————	————	————
(b) Temperature of the reaction mixture in °C	————	————	————	————
(c) Relative rate = 1000/t	————	————	————	————
(d) Log$_{10}$ of relative rate	————	————	————	————
(e) Temperature T in K	————	————	————	————
(f) $1/T$, K^{-1}	————	————	————	————

To evaluate E_a, make a graph of log relative rate vs. $1/T$ on the graph paper provided. (See Appendix V).

(continued on following page)

(continued)

Find the slope of the line obtained by drawing the best straight line through the experimental points. Make the scale such that each square on the abscissa equals 0.00005 and each square on the ordinate equals 0.2.

Slope = _____

The slope of the line equals $-E_a/2.3R$, where $R = 8.31$ joules/mole K if E_a is to be in joules per mole. Find the activation energy, E_a, for the reaction.

E_a = _____ joules

C. Effect of a Catalyst on Reaction Rate (Optional)

	Reaction 1	Catalyzed Reaction 1
Time for color to appear (seconds)	_____	_____

Would you expect the activation energy, E_a, for the catalyzed reaction to be greater than, less than, or equal to the activation energy for the uncatalyzed reaction? Why?

Name _____ Section _____

RATES OF CHEMICAL REACTIONS,
II. A CLOCK REACTION
(Data and Calculations)

Advance Study Assignment: Rate of Reactions, II. A Clock Reaction

1. A student studied the clock reaction described in this experiment. She set up Reaction Mixture 3 by mixing 10 mL 0.010 M KI, 10 mL 0.0010 M $Na_2S_2O_3$, 20 mL 0.040 M $KBrO_3$, and 10 mL 0.10 M HCl using the procedure given. It took about 40 seconds for the color to turn blue.

   a. She found the concentrations of each reactant in the reacting mixture by realizing that the number of moles of each reactant did not change when that reactant was mixed with the others, but that its concentration did. For any reactant A,

   $$\text{no. moles A} = M_{A\ stock} \times V_{stock} = M_{A\ mixture} \times V_{mixture}$$

   The volume of the mixture was 50 mL. Revising the equation, she obtained

   $$M_{A\ mixture} = M_{A\ stock} \times \frac{V_{stock}(mL)}{50\ mL}$$

   Find the concentrations of each reactant by using the above equation.

   $(I^-) =$ _____ M; $(BrO_3^-) =$ _____ M; $(H^+) =$ _____ M

   b. What was the relative rate of the reaction $(1000/t)$? _____

   c. Knowing the relative rate of reaction for Mixture 3 and the concentrations of I^-, BrO_3^-, and H^+ in that mixture, she was able to set up Equation 5 for the relative rate of the reaction. The only quantities that remained unknown were k', m, n, and p. Set up Equation 5 as she did, presuming she did it properly.

2. For Reaction Mixture 1 the student found that 75 seconds were required. On dividing Equation 5 for Reaction Mixture 1 by Equation 5 for Reaction Mixture 3, and after canceling out the common terms (k', terms in (I^-) and (H^+)), she got the following equation:

   $$\frac{13.3}{25} = \left(\frac{0.0040}{0.0080}\right)^n = \left(\frac{1}{2}\right)^n$$

   Recognizing that 13.3/25 is about equal to 1/2, she obtained an approximate value for n. What was that value?

   $n =$ _____

(continued on following page)

(continued)

By taking logarithms of both sides of the equation, she got an exact value for n. What was that value?

$$n = \underline{\hspace{2cm}}$$

Since orders of reactions are often integers, she reported her approximate value as the order of the reaction with respect to BrO_3^-.

Experiment 20

PROPERTIES OF SYSTEMS IN CHEMICAL EQUILIBRIUM—LE CHÂTELIER'S PRINCIPLE

When working in the laboratory, one often makes observations that at first sight are surprising and hard to explain. One might add a reagent to a solution and obtain a precipitate. Addition of more of that reagent to the precipitate causes it to dissolve. A violet solution turns yellow on addition of a reagent. Subsequent addition of another reagent brings back first a green solution and then the original violet one. Clearly, chemical reactions are occurring, but how and why they behave as they do is not at once obvious.

In this experiment we will examine and attempt to explain several observations of the sort we have mentioned. Central to our explanation will be recognition of the fact that chemical systems tend to exist in a state of equilibrium. If one disturbs the equilibrium in one way or another, the reaction may shift to the left or right, producing the kinds of effects we have mentioned. If one can understand the principles governing the equilibrium system, it is often possible to see how one might disturb the system, such as by adding a particular reagent or heat, and so cause it to change in a desirable way.

Before proceeding to specific examples, let us examine the situation in a general way, noting the key principle that allows us to make a system in equilibrium behave as we wish. Consider the reaction

$$A(aq) \rightleftharpoons B(aq) + C(aq) \tag{1}$$

where A, B, and C are molecules or ions in solution. If we have a mixture of these species in equilibrium, it turns out that their concentrations are not completely unrelated. Rather, there is a condition that those concentrations must meet, namely that

$$\frac{[B] \times [C]}{[A]} = K_c \tag{2}$$

where K_c is a constant, called the equilibrium constant for the reaction. For a given reaction at any given temperature, K_c has a particular value.

When we say that K_c has a particular value, we mean just that. For example, we might find that, for a given solution in which Reaction 1 can occur, when we substitute the equilibrium values for the molarities of A, B, and C into Equation 2, we get a value of 10 for K_c. Now, suppose we add more of species A to that solution. What will happen? Remember, K_c can't change. If we substitute the new higher molarity of A into Equation 2 we get a value that is smaller than K_c. This means that the system is not in equilibrium, and *must* change in some way to get back to equilibrium. How can it do this? It can do this by shifting to the right, producing more B and C and using up some A. It *must* do this, and *will,* until the molarities of C, B, and A reach values that, on substitution into Equation 2, equal 10. At that point the system is once again in equilibrium. In the new equilibrium state, [B] and [C] are greater than they were initially, and [A] is larger than its initial value but smaller than if there had been no forced shift to the right.

The conclusion you should reach on reading the last paragraph is that *one can always cause a reaction to shift to the right by increasing the concentration of a reactant.* An *increase* in concentration of a *product* will force a *shift* to the *left.* By a similar argument we find that a *decrease* in *reactant* concentration causes a *shift* to the *left; a decrease* in *product* concentration produces a *shift* to the *right.* This is all true because K_c does not change (unless you change the temperature). The changes in concentration that one can produce by adding particular reagents may be simply enormous, so the shifts in the equilibrium system may also be enormous. Much of the mystery of chemical behavior disappears once you understand this idea.

Another way one might disturb an equilibrium system is by changing its temperature. When this happens, the value of K_c changes. It turns out that the change in K_c depends upon the enthalpy change, ΔH, for the reaction. If ΔH is positive, greater than zero (endothermic reaction), K_c increases with increasing T. If ΔH is negative (exothermic reaction), K_c decreases with an increase in T. Let us return to our original equilibrium between A, B, and C, where K_c equals 10. Let us assume that ΔH for Reaction 1 is -40 kJ. If we raise the temperature, K_c will go down ($\Delta H < 0$), say to a value of 1. This means that the system will no longer be in equilibrium. Substitution of the initial values of [A], [B], and [C] into Equation 2 produces a value that is too big, 10 instead of 1. How can the system change itself to regain equilibrium? It must of necessity shift to the left, lowering [B] and [C] and raising [A]. This will make the expression in Equation 2 smaller. The shift will continue until the concentrations of A, B, and C, on substitution into Equation 2, give the expression a value of 1.

From the discussion in the previous paragraph, you should be able to conclude that an equilibrium system will shift to the left on being heated if the reaction is exothermic ($\Delta H < 0$, K_c goes down). It will shift to the right if the reaction is endothermic ($\Delta H > 0$, K_c goes up). Again, since we can change temperatures very markedly, we can shift equilibria a long, long way. An endothermic reaction that at 25°C has an equilibrium state that consists mainly of reactants might at 1000°C exist almost completely as products.

The effects of concentration and temperature on systems in chemical equilibrium are often summarized by Le Châtelier's principle. The principle states that:

If you attempt to change a system in chemical equilibrium, it will react in such a way as to counteract the change you attempted.

If you think about the principle for a while, you will see that it predicts the same kind of behavior as we did by using the properties of K_c. Increasing the concentration of a reactant will, by the principle, cause a change that decreases that concentration; that change must be a shift to the right. Increasing the temperature of a reaction mixture will cause a change that tends to absorb heat; that change must be a shift in the endothermic direction. The principle is an interesting one, but does require more careful reasoning in some cases than the more direct approach we employed. For the most part we will find it more useful to base our arguments on the properties of K_c.

In working with aqueous systems, the most important equilibrium is often that which involves the dissociation of water into H^+ and OH^- ions:

$$H_2O \rightleftharpoons H^+(aq) + OH^-(aq) \qquad K_c = [H^+][OH^-] = 1 \times 10^{-14} \qquad (3)$$

In this reaction the concentration of water is very high and is essentially constant at about 55 M; it is incorporated into K_c. The value of K_c is very small, which means that in **any** water system the product of [H^+] and [OH^-] must be very small. In pure water, [H^+] equals [OH^-] equals 1×10^{-7} M.

Although the product, [H^+] $\times$ [OH^-] is small, that does not mean that both concentrations are necessarily small. If, for example, we dissolve HCl in water, the HCl in the solution will dissociate completely to H^+ and Cl^- ions; in 1 M HCl, [H^+] will become 1 M, and there is nothing that Reaction 3 can do about changing that concentration appreciably. Rather, Reaction 3 must occur in such a direction as to maintain equilibrium. It does this by lowering $[OH^-]$ by reaction to the left; this uses up a little bit of H^+ ion and drives $[OH^-]$ to the value it must have when $[H^+]$ is 1 M, namely, 1×10^{-14} M. In 1 M HCl, $[OH^-]$ is a factor of ten million *smaller* than it is in water. This makes the properties of 1 M HCl quite different from those of water, particularly where H^+ and OH^- ions are involved.

If we take 1 M HCl and add a solution of NaOH to it, an interesting situation develops. Like HCl, NaOH is completely dissociated in solution, so in 1 M NaOH, $[OH^-]$ is equal to 1 M. If we add 1 M NaOH to 1 M HCl, we will initially raise $[OH^-]$ ions way above 1×10^{-14} M. However, Reaction 3 cannot be in equilibrium when both $[H^+]$ and $[OH^-]$ are high; reaction must occur to re-establish equilibrium. The added OH^- ions react with H^+ ions to form H_2O, decreasing both concentrations until equilibrium is established. If only a small amount of OH^- ion is added, it will essentially all be used up; $[H^+]$ will remain high, and $[OH^-]$ will still be very small, but somewhat larger than 10^{-14} M. If we add OH^- ion until the amount added equals in moles the amount of H^+ originally present, then Reaction 3 will go to the left until $[H^+]$ equals $[OH^-]$ equals 1×10^{-7} M, and both concentrations will be very small. Further addition of OH^- ion will raise $[OH^-]$ to much higher values, easily as high as 1 M. In such a solution, $[H^+]$ would be very low, 1×10^{-14} M. So, in aqueous solution, depending on the solutes present, we can have $[H^+]$ and $[OH^-]$ range from about 1 M to 10^{-14} M, or 14 orders of magnitude. This will have a tremendous effect on **any** other equilibrium system in which $[H^+]$ or $[OH^-]$ ions are reactants. Similar situations arise in other equilibrium systems in which the concentration of a reactant or product can be changed significantly by adding a particular reagent.

In many equilibrium systems, several equilibria are present simultaneously. For example, in aqueous solution, Reaction 3 must **always** be in equilibrium. There may, in addition, be equilibria between the solutes in the aqueous solution. Some examples are those in Reactions 4, 5, 7, 8, 9, and 10 in the Experimental Procedure section. In some of those reactions, H^+ and OH^- ions appear; in others, they do not. In Reaction 4, for example, H^+ ion is a product. The molarity of H^+ in Reaction 4 is *not* determined by the indicator HMV, since it is only present in a tiny amount. Reaction 4 will have an equilibrium state that is fixed by the state of Reaction 3, which as we have seen depends markedly on the presence of solutes such as HCl or NaOH. Reactions 8 and 9 can similarly be controlled by Reaction 3. Reactions 5 and 7, which do not involve H^+ or OH^- ions, are not dependent on Reaction 3 for their equilibrium state. Reaction 10 is sensitive to NH_3 concentration and can be driven far to the right by addition of a reagent such as 6 M NH_3.

Experimental Procedure

In this experiment we will work with several equilibrium systems, each of which is similar to the A-B-C system we discussed. We will alter these systems in various ways, forcing shifts to the right and left by changing concentrations and temperature. You will be asked to interpret your observations in terms of the principles we have presented.

A. Acid-Base Indicators

There is a large group of chemical substances, called acid-base indicators, which change color in solution when $[H^+]$ changes. A typical substance of this sort is called methyl violet, which we will give the formula HMV. In solution HMV dissociates as follows:

$$HMV(aq) \rightleftharpoons H^+(aq) + MV^-(aq) \qquad (4)$$
$$\text{yellow} \qquad\qquad\qquad \text{violet}$$

HMV has an intense yellow color, while the anion MV^- is violet. The color of the indicator in solution depends very strongly on $[H^+]$.

Step 1. Add about 5 mL of distilled water to a regular test tube. Add a few drops of methyl violet indicator. Report the color of the solution on the Data page.

Step 2. How could you force the equilibrium system to go to the other form (color)? Select a reagent that should do this and add it to the solution, drop by drop, until the color change is complete. If your reagent works, write its formula on the Data page. If it doesn't, try another until you find one that does. Work with 6 M reagents if they are available.

Step 3. Equilibrium systems are reversible. That is, the reaction can be driven to the left and right many times by changing the conditions in the system. How can you force the system in Step 2 to revert to its original color? Select a reagent that should do this and add it drop by drop until the color has become the original one. Again, if your first choice was incorrect, try another reagent. On the Data page write the formula of the reagent that was effective. Answer all the questions for Part A before going on to Part B.

B. Solubility Equilibrium; Finding a Value for K_{sp}

Many ionic substances have limited water solubility. A typical example is $PbCl_2$, which dissolves to some extent in water according to the reaction

$$PbCl_2(s) \rightleftharpoons Pb^{2+}(aq) + 2\,Cl^-(aq) \tag{5}$$

The equilibrium constant for this reaction takes the form

$$K_c = [Pb^{2+}] \times [Cl^-]^2 = K_{sp} \tag{6}$$

The $PbCl_2$ does not enter into the expression because it is a solid, and so has a constant effect on the system, independent of its amount. The equilibrium constant for a solubility equilibrium is called the solubility product, and is given the symbol K_{sp}.

For the equilibrium in Reaction 5 to exist, there *must* be some solid $PbCl_2$ present in the system. If there is no solid, there is no equilibrium; Equation 6 is not obeyed, and $[Pb^{2+}] \times [Cl^-]^2$ must be *less* than the value of K_{sp}. If the solid is present, even in a tiny amount, then the values of $[Pb^{2+}]$ and $[Cl^-]$ are subject to Equation 6.

Step 1. Set up a hot-water bath, using a 400-mL beaker half full of water. Start heating the water with a burner while proceeding with Step 2.

Step 2. To a regular test tube add 5.0 mL 0.30 M $Pb(NO_3)_2$. In this solution $[Pb^{2+}]$ equals 0.30 M. Add 5.0 mL 0.30 M HCl to a 10-mL graduated cylinder. In this solution $[Cl^-]$ equals 0.30 M. Add 1 mL of the HCl solution to the $Pb(NO_3)_2$ solution. Stir, and wait about 15 seconds. What happens? Enter your result.

Step 3. To the $Pb(NO_3)_2$ solution add the HCl in 1-mL increments until a noticeable amount of white solid $PbCl_2$ is present after stirring. Record the total volume of HCl added at that point.

Step 4. Put the test tube with the precipitate of $PbCl_2$ into the hot-water bath. Stir for a few moments. What happens? Enter your observations. Cool the test tube under the cold-water tap. What happens?

Step 5. Rinse out your graduated cylinder and then add about 5.0 mL of distilled water to the cylinder. Add water in 1-mL increments to the mixture in the test tube, stirring well after each addition. Record the volume of water added when the precipitate just dissolves. Answer the questions and do the calculations in Part B before proceeding.

C. Complex Ion Equilibria

Many metallic ions in solution exist not as simple ions but, rather, as complex ions in combination with other ions or molecules, called ligands. For example, the Co^{2+} ion in solution exists as the pink $Co(H_2O)_6^{2+}$ complex ion, and Cu^{2+} as the blue $Cu(H_2O)_4^{2+}$ complex ion. In both of these ions the ligands are H_2O molecules. Complex ions are reasonably stable but may be converted to other complex ions on addition of ligands that form more stable complexes than the original ones. Among the common ligands that may form complex species, OH^-, NH_3, and Cl^- are important.

An interesting Co(II) complex is the $CoCl_4^{2-}$ ion, which is blue. This ion is stable in concentrated Cl^- solutions. Depending upon conditions, Co(II) in solution may exist as either $Co(H_2O)_6^{2+}$ or as $CoCl_4^{2-}$. The principles of chemical equilibrium can be used to predict which ion will be present:

$$Co(H_2O)_6^{2+}(aq) + 4\ Cl^-(aq) \rightleftharpoons CoCl_4^{2-}(aq) + 6\ H_2O \qquad (7)$$

Step 1. Put a few small crystals (~ 0.1 g) of $CoCl_2 \cdot 6\ H_2O$ in a regular test tube. Add 2 mL of 12 M HCl (**CAUTION**) and stir to dissolve the crystals. Record the color of the solution.

Step 2. Add 2-mL portions of distilled water, stirring after each dilution, until no further color change occurs. Record the new color.

Step 3. Place the test tube into the hot-water bath and note any change in color. Cool the tube under the water tap and report your observations. Complete the questions in Part C before continuing.

D. Dissolving Insoluble Solids

We saw in Part B that we can dissolve more $PbCl_2$ by either heating its saturated solution or by simply adding water. In most cases these procedures won't work very well on other solids because they are typically much less soluble than $PbCl_2$.

There are, however, some very powerful methods for dissolving solids that depend for their effectiveness upon the principles of equilibrium. As an example of an insoluble substance, we might consider $Zn(OH)_2$:

$$Zn(OH)_2(s) \rightleftharpoons Zn^{2+}(aq) + 2\ OH^-(aq) \qquad K_{sp} = 5 \times 10^{-17} = [Zn^{2+}] \times [OH^-]^2 \quad (8)$$

The equilibrium constant for Reaction 8 is very small, which tells us that the reaction does not go very far to the right or, equivalently, that $Zn(OH)_2$ is almost completely insoluble in water. Adding a few drops of a solution containing OH^- ion to one containing Zn^{2+} ion will cause precipitation of $Zn(OH)_2$.

At first sight you might well wonder how one could possibly dissolve, say, 1 mole of $Zn(OH)_2$ in an aqueous solution. If, however, you examine Equation 8, you can see, from the equation for K_{sp}, that in the saturated solution $[Zn^{2+}] \times [OH^-]^2$ must equal 5×10^{-17}. If, by some means, we can lower that product to a value *below* 5×10^{-17}, then $Zn(OH)_2$ will dissolve, until the product becomes equal to K_{sp}, where equilibrium will again exist. To lower the product, we need to lower the concentration of either Zn^{2+} or OH^- very drastically. This turns out to be very easy to do. To lower $[OH^-]$ we can add H^+ ions from an acid. If we do that, we drive Reaction 3 to the left, making $[OH^-]$ very, very small—small enough to dissolve substantial amounts of $Zn(OH)_2$.

To lower $[Zn^{2+}]$ we can take advantage of the fact that zinc(II) forms stable complex ions with both OH^- and NH_3:

$$Zn^{2+}(aq) + 4\ OH^-(aq) \rightleftharpoons Zn(OH)_4^{2-}(aq) \qquad K_1 = 3 \times 10^{15} \qquad (9)$$

$$Zn^{2+}(aq) + 4\ NH_3(aq) \rightleftharpoons Zn(NH_3)_4^{2+}(aq) \qquad K_2 = 1 \times 10^9 \qquad (10)$$

In high concentrations of OH^- ion, Reaction 9 is driven strongly to the right, making $[Zn^{2+}]$ very low. The same thing would happen in solutions containing high concentrations of NH_3. In both media we would therefore expect that $Zn(OH)_2$ might dissolve, since if $[Zn^{2+}]$ is very low, Reaction 8 must go to the right.

Step 1. To each of three small test tubes add about 2 mL 0.1 M $Zn(NO_3)_2$. In this solution $[Zn^{2+}]$ equals 0.1 M. To each test tube add one drop 6 M NaOH and stir. Report your observations.

Step 2. To the first tube add 6 M HCl drop by drop, with stirring. To the second add 6 M NaOH, again drop by drop. To the third add 6 M NH_3. Note what happens in each case.

Step 3. Repeat Steps 1 and 2, this time using a solution of 0.1 M $Mg(NO_3)_2$. Record your observations. Answer the questions in Part D.

The residues from Parts B and C should be poured in the waste crock. Those from Parts A and D may be poured down the sink.

Data and Observations: Systems in Chemical Equilibrium

A.   Acid-Base Indicators

1. Color of methyl violet in water _____

2. Reagent causing color change _____

3. Reagent causing shift back _____. Explain, by considering how changes in $[H^+]$ will cause Reaction 4 to shift to right and left, why the reagents in Steps 2 and 3 caused the solution to change color. Note that Reactions 3 and 4 must both go to equilibrium after a reagent is added.

B.   Solubility Equilibrium; Finding a Value for K_{sp}

2. Vol. 0.30 M $Pb(NO_3)_2$ = 5.0 mL   No. moles Pb^{2+}(M × V(lit)) = 1.5×10^{-3} moles

Observations:

3. Vol. 0.30 M HCl used _____ mL   No. moles Cl^- added _____ moles

4. Observations: in hot water _____   in cold water _____

5. Volume H_2O added to dissolve $PbCl_2$ _____ mL

Total volume of solution _____ mL

a. Explain why $PbCl_2$ did not precipitate immediately on addition of HCl. (What condition must be met by $[Pb^{2+}]$ and $[Cl^-]$ if $PbCl_2$ is to form?)

b. Explain your observations in Step 4. (In which direction did Reaction 5 shift when heated? What must have happened to the value of K_{sp} in the hot solution? What does this tell you about the sign of ΔH in Reaction 5?)

c. Explain why the $PbCl_2$ dissolved when water was added in Step 5. (What was the effect of the added water on $[Pb^{2+}]$ and $[Cl^-]$? In what direction would such a change drive Reaction 5?)

(continued on following page)

(continued)

    d. Given the numbers of moles of Pb^{2+} and Cl^- in the final solution in Step 5, and the volume of that solution, calculate $[Pb^{2+}]$ and $[Cl^-]$ in that solution.

$$[Pb^{2+}] \underline{\hspace{2cm}} M; [Cl^-] \underline{\hspace{2cm}} M$$

Noting that the molarities just calculated are essentially those in equilibrium with solid $PbCl_2$, calculate $[Pb^{2+}] \times [Cl^-]^2$. This is equal to K_{sp} for $PbCl_2$.

$$K_{sp} = \underline{\hspace{2cm}}$$

C. Complex Ion Equilibria

  1. Color of $CoCl_2 \cdot 6 H_2O$ _____

    Color in solution in 12 M HCl   _____

  2. Color in diluted solution   _____

  3. Color of hot solution   _____

    Color of cooled solution   _____

    Formula of Co(II) complex ion present in solution in

    a. 12 M HCl   _____

    b. diluted solution   _____

    c. hot solution   _____

    d. cooled solution   _____

    Explain the color change that occurred when

    a. water was added in Step 2. (Consider how a change in $[Cl^-]$ and $[H_2O]$ will shift Reaction 7.)

    b. the diluted solution was heated. (How would Reaction 7 shift if K_c went up? How did increasing the temperature affect the value of K_c? What is the sign of ΔH in Reaction 7?)

(continued)

D. Dissolving Insoluble Solids

1. Observations on addition of a drop of 6 M NaOH to $Zn(NO_3)_2$ solution:

2. Effect on solubility of $Zn(OH)_2$:

   a. of added HCl solution

   b. of added NaOH solution

   c. of added NH_3 solution

3. Observations on addition of one drop of 6 M NaOH to $Mg(NO_3)_2$ solution:

   Effect on solubility of $Mg(OH)_2$:

   a. of added HCl solution

   b. of added NaOH solution

   c. of added NH_3 solution

Explain your observations in Step 1. (Consider Reaction 8; how is it affected by addition of OH^- ion?)

In Step 2a, how does an increase in $[H^+]$ affect Reaction 3? (What does that do to Reaction 8?) Explain your observations in Step 2a.

In Step 2b, how does an increase in $[OH^-]$ affect Reaction 9? What does that do to Reaction 8? Explain your observations in Step 2b.

In Step 2c, how does an increase in $[NH_3]$ affect Reaction 10? What does that do to Reaction 8? Explain your observations in Step 2c.

In Step 3, you probably found that $Mg(OH)_2$ was similar in some ways in its behavior to that of $Zn(OH)_2$, but different in others.

a. How was it similar? Explain that similarity. (In particular, why would any insoluble hydroxide tend to dissolve in acidic solution?)

b. How was it different? Explain that difference. (In particular, does Mg^{2+} appear to form complex ions with OH^- and NH_3? What would we observe if it did? If it did not?)

Advance Study Assignment: Systems in Chemical Equilibrium

1. Methyl orange, HMO, is a common acid-base indicator. In solution it ionizes according to the equation:

$$HMO(aq) \rightleftharpoons H^+(aq) + MO^-(aq)$$
$$\text{red} \qquad\qquad\qquad \text{yellow}$$

If methyl orange is added to distilled water, the solution turns yellow. If a drop or two of 6 M HCl is added to the yellow solution, it turns red. If to that solution one adds a few drops of 6 M NaOH, the color reverts to yellow.

  a. Why does adding 6 M HCl to the yellow solution of methyl orange tend to cause the color to change to red? (Note that in solution HCl exists as H^+ and Cl^- ions.)

  b. Why does adding 6 M NaOH to the red solution tend to make it turn back to yellow? (Note that in solution NaOH exists as Na^+ and OH^- ions. How does increasing $[OH^-]$ shift Reaction 3 in the discussion section? How would the resulting change in $[H^+]$ affect the dissociation reaction of HMO?)

2. Zinc hydroxide is only very slightly soluble in water. The reaction by which it goes into solution is:

$$Zn(OH)_2(s) \rightleftharpoons Zn^{2+}(aq) + 2\ OH^-(aq)$$

  a. Formulate the expression for the equilibrium constant, K_{sp}, for the above reaction.

  b. It is possible to dissolve significant amounts of $Zn(OH)_2$ in solutions in which the concentration of either Zn^{2+} or OH^- is very, very small. Explain, using K_{sp}, why this is the case.

  c. Explain why $Zn(OH)_2$ might have very appreciable solubility in 1 M HCl. (Consider the effect of Reaction 3 on the $Zn(OH)_2$ solution reaction.)

DETERMINATION OF THE EQUILIBRIUM CONSTANT FOR A CHEMICAL REACTION

When chemical substances react, the reaction typically does not go to completion. Rather, the system goes to some intermediate state in which both the reactants and products have concentrations that do not change with time. Such a system is said to be in chemical equilibrium. When in equilibrium at a particular temperature, a reaction mixture obeys the Law of Chemical Equilibrium, which imposes a condition on the concentrations of reactants and products. This condition is expressed in the equilibrium constant K_c for the reaction.

In this experiment we will study the equilibrium properties of the reaction between iron(III) ion and thiocyanate ion:

$$Fe^{3+}(aq) + SCN^-(aq) \rightleftharpoons FeSCN^{2+}(aq) \tag{1}$$

When solutions containing Fe^{3+} ion and thiocyanate ion are mixed, Reaction 1 occurs to some extent, forming the $FeSCN^{2+}$ complex ion, which has a deep red color. As a result of the reaction, the equilibrium amounts of Fe^{3+} and SCN^- will be less than they would have been if no reaction occurred; for every mole of $FeSCN^{2+}$ that is formed, one mole of Fe^{3+} and one mole of SCN^- will react. According to the Law of Chemical Equilibrium, the equilibrium constant expression K_c for Reaction 1 is formulated as follows:

$$\frac{[FeSCN^{2+}]}{[Fe^{3+}][SCN^-]} = K_c \tag{2}$$

The value of K_c in Equation 2 is constant at a given temperature. This means that mixtures containing Fe^{3+} and SCN^- will react until Equation 2 is satisfied, so that the same value of the K_c will be obtained no matter what initial amounts of Fe^{3+} and SCN^- were used. Our purpose in this experiment will be to find K_c for this reaction for several mixtures made up in different ways, and to show that K_c indeed has the same value in each of the mixtures. The reaction is a particularly good one to study because K_c is of a convenient magnitude and the color of the $FeSCN^{2+}$ ion makes for an easy analysis of the equilibrium mixture.

The mixtures will be prepared by mixing solutions containing known concentrations of iron(III) nitrate, $Fe(NO_3)_3$, and potassium thiocyanate, KSCN. The color of the $FeSCN^{2+}$ ion formed will allow us to determine its equilibrium concentration. Knowing the initial composition of a mixture and the equilibrium concentration of $FeSCN^{2+}$, we can calculate the equilibrium concentrations of the rest of the pertinent species and then determine K_c.

Since the calculations required in this experiment may not be apparent, we will go through a step-by-step procedure by which they can be made. As a specific example, let us assume that we prepare a mixture by mixing 10.0 mL of 2.00×10^{-3} M $Fe(NO_3)_3$ with 10.0 mL of 2.00×10^{-3} M KSCN. As a result of Reaction 1, some red $FeSCN^{2+}$ ion is formed. By the method of analysis described later, its concentration at equilibrium is found to be 1.50×10^{-4} M. Our problem is to find K_c for the reaction from this information. To do this we first need to find the initial number of moles of each reactant in the mixture. Second, we determine the number of moles of product that were formed at equilibrium. Since the product was formed at the expense of reactants, we can calculate the amount of each reactant that was used up. In the third step we find the number of moles of each reactant remaining in the equilibrium mixture. Fourth, we determine the concentration of each reactant. Finally, in the fifth step we evaluate K_c for the reaction.

Step 1.   Finding the Initial Number of Moles of Each Reactant. This requires relating the volumes and concentrations of the reagent solutions that were mixed to the numbers of moles of each reactant species in those solutions. By the definition of the molarity, M_A, of a species A,

$$M_A = \frac{\text{no. moles } A}{\text{no. liters of solution, } V} \quad \text{or} \quad \text{no. moles } A = M_A \times V \quad (3)$$

Using Equation 3, it is easy to find the initial number of moles of Fe^{3+} and SCN^-. For each solution the volume used was 10.0 mL, or 0.0100 L. The molarity of each of the solutions was 2.00×10^{-3} M, so $M_{Fe^{3+}} = 2.00 \times 10^{-3}$ M and $M_{SCN^-} = 2.00 \times 10^{-3}$ M. Therefore, in the reagent solutions, we find that

initial no. moles $Fe^{3+} = M_{Fe^{3+}} \times V = 2.00 \times 10^{-3}$ M $\times 0.0100$ L $= 20.0 \times 10^{-6}$ moles

initial no. moles $SCN^- = M_{SCN^-} \times V = 2.00 \times 10^{-3}$ M $\times 0.0100$ L $= 20.0 \times 10^{-6}$ moles

Step 2.   Finding the Number of Moles of Product Formed. Here again we can use Equation 3 to advantage. The concentration of $FeSCN^{2+}$ was found to be 1.50×10^{-4} M at equilibrium. The volume of the mixture at equilibrium is the *sum* of the two volumes that were mixed, and is 20.0 mL, or 0.0200 L. So,

no. moles $FeSCN^{2+} = M_{FeSCN^{2+}} \times V = 1.50 \times 10^{-4}$ M $\times 0.0200$ L $= 3.00 \times 10^{-6}$ moles

The number of moles of Fe^{3+} and SCN^- that were *used up* in producing the $FeSCN^{2+}$ must also both be equal to 3.00×10^{-6} moles since, by Equation 1, it takes *one mole* Fe^{3+} *and one mole* SCN^- to make each mole of $FeSCN^{2+}$.

Step 3.   Finding the Number of Moles of Each Reactant Present at Equilibrium. In Step 1 we determined that initially we had 20.0×10^{-6} moles Fe^{3+} and 20.0×10^{-6} moles SCN^- present. In Step 2 we found that in the reaction 3.00×10^{-6} moles Fe^{3+} and 3.00×10^{-6} moles SCN^- were used up. The number of moles present at equilibrium must equal the number we started with minus the number that reacted. Therefore, *at equilibrium,*

no. moles at equilibrium = initial no. moles − no. moles used up

equil. no. moles $Fe^{3+} = 20.0 \times 10^{-6} - 3.00 \times 10^{-6} = 17.0 \times 10^{-6}$ moles   (4)

equil. no. moles $SCN^- = 20.0 \times 10^{-6} - 3.00 \times 10^{-6} = 17.0 \times 10^{-6}$ moles

Step 4.   Find the Concentrations of All Species at Equilibrium. Experimentally, we obtained the concentration of $FeSCN^{2+}$ directly. $[FeSCN^{2+}] = 1.50 \times 10^{-4}$ M. The concentrations of Fe^{3+} and SCN^- follow from Equation 3. The number of moles of each of these species at equilibrium was obtained in Step 3. The volume of the mixture being studied was 20.0 mL, or 0.0200 L. So, *at equilibrium,*

$$[Fe^{3+}] = M_{Fe^{3+}} = \frac{\text{no. moles } Fe^{3+}}{\text{volume of solution}} = \frac{17.0 \times 10^{-6} \text{ moles}}{0.0200 \text{ L}} = 8.50 \times 10^{-4} \text{ M}$$

$$[SCN^-] = M_{SCN^-} = \frac{\text{no. moles } SCN^-}{\text{volume of solution}} = \frac{17.0 \times 10^{-6} \text{ moles}}{0.0200 \text{ L}} = 8.50 \times 10^{-4} \text{ M}$$

Step 5. Finding the Value of K_c for the Reaction. Once the equilibrium concentrations of all the reactants and products are known, one needs merely to substitute into Equation 2 to determine K_c:

$$K_c = \frac{[FeSCN^{2+}]}{[Fe^{3+}][SCN^-]} = \frac{1.50 \times 10^{-4}}{(8.50 \times 10^{-4}) \times (8.50 \times 10^{-4})} = 208$$

In this experiment you will obtain data similar to that shown in this example. The calculations involved in processing that data are completely analogous to those we have made. (Actually, your results will differ from the ones we obtained, since the data in our example were obtained at a different temperature and so relate to a different value of K_c.)

In carrying out this analysis we made the assumption that the reaction which occurred was given by Equation 1. There is no inherent reason why the reaction might not have been

$$Fe^{3+}(aq) + 2\ SCN^-(aq) \rightleftharpoons Fe(SCN)_2{}^+(aq) \tag{5}$$

If you are interested in matters of this sort, you might ask how we know whether we are actually observing Reaction 1 or Reaction 5. The line of reasoning is the following. If Reaction 1 is occurring, K_c for that reaction as we calculate it should remain constant with different reagent mixtures. If, however, Reaction 5 is going on, K_c as calculated for that reaction should remain constant. In the optional part of the Calculations section, we will assume that Reaction 5 occurs and make the analysis of K_c on that basis. The results of the two sets of calculations should make it clear that Reaction 1 is the one that we are studying.

Two analytical methods can be used to determine $[FeSCN^{2+}]$ in the equilibrium mixtures. The more precise method uses a spectrophotometer, which measures the amount of light absorbed by the red complex at 447 nm, the wavelength at which the complex most strongly absorbs. The absorbance, A, of the complex is proportional to its concentration, M, and can be measured directly on the spectrophotometer:

$$A = kM \tag{6}$$

Your instructor will show you how to operate the spectrophotometer, if available to your laboratory, and will provide you with a calibration curve or equation from which you can find $[FeSCN^{2+}]$ once you have determined the absorbance of your solutions. See Appendix IV for information about spectrophotometers.

In the other analytical method a solution of known concentration of $FeSCN^{2+}$ is prepared. The $[FeSCN^{2+}]$ concentrations in the solutions being studied are found by comparing the color intensities of these solutions with that of the known. The method involves matching the color intensity of a given depth of unknown solution with that for an adjusted depth of known solution. The actual procedure and method of calculation are discussed in the Experimental Procedure section.

In preparing the mixtures in this experiment we will maintain the concentration of H^+ ion at 0.5 M. The hydrogen ion does not participate directly in the reaction, but its presence is necessary to avoid the formation of brown-colored species such as $FeOH^{2+}$, which would interfere with the analysis of $[FeSCN^{2+}]$.

Experimental Procedure

Label five regular test tubes 1 to 5, with labels or by noting their positions on your test tube rack. Pour about 30 mL 2.00×10^{-3} M $Fe(NO_3)_3$ in 1 M HNO_3 into a dry 100-mL beaker.

Pipet 5.00 mL of that solution into each test tube. Then add about 20 mL 2.00×10^{-3} M KSCN to another dry 100-mL beaker. Pipet 1, 2, 3, 4, and 5 mL from the KSCN beaker into each of the corresponding test tubes labeled 1 to 5. Then pipet the proper number of milliliters of water into each test tube to bring the total volume in each tube to 10.00 mL. The volumes of reagents to be added to each tube are summarized in Table 21.1. See Appendix IV for a discussion of the use of pipets.

Table 21.1

	Test Tube No.				
	1	*2*	*3*	*4*	*5*
Volume $Fe(NO_3)_3$ solution (mL)	5.00	5.00	5.00	5.00	5.00
Volume KSCN solution (mL)	1.00	2.00	3.00	4.00	5.00
Volume H_2O (ml)	4.00	3.00	2.00	1.00	0.00

Mix each solution thoroughly with a glass stirring rod. Be sure to dry the stirring rod after mixing each solution.

Method I. Analysis by Spectrophotometric Measurement

Place a portion of the mixture in tube 1 in a spectrophotometer cell, as demonstrated by your instructor, and measure the absorbance of the solution at 447 nm. Determine the concentration of $FeSCN^{2+}$ from the calibration curve provided for each instrument or from the equation furnished to you. Record the value on the Data page. Repeat the measurement using the mixtures in each of the other test tubes. For a discussion of how absorbance and concentration are related, see Appendix IV.

Method II. Analysis by Comparison with a Standard

Prepare a solution of known $[FeSCN^{2+}]$ by pipetting 10.00 mL 0.200 M $Fe(NO_3)_3$ in 1 M HNO_3 into a test tube and adding 2.00 mL 0.00200 M KSCN and 8.00 mL water. Mix the solution thoroughly with a stirring rod.

Since in this solution $[Fe^{3+}] \gg [SCN^-]$, Reaction 1 is driven strongly to the right. You can assume without serious error that essentially all the SCN^- added is converted to $FeSCN^{2+}$. Assuming that this is the case, calculate $[FeSCN^{2+}]$ in the standard solution and record the value on the Data page.

The $[FeSCN^{2+}]$ in the unknown mixture in test tubes 1 to 5 can be found by comparing the intensity of the red color in these mixtures with that of the standard solution. This can be done by placing the test tube containing mixture 1 next to a test tube containing the standard. Look down both test tubes toward a well-illuminated piece of white paper on the laboratory bench.

Pour out the standard solution into a dry, clean beaker until the color intensity you see down the tube containing the standard matches that which you see when looking down the tube containing the unknown. Use a well-lit piece of white paper as your background. When the colors match, the following relation is valid:

$[FeSCN^{2+}]_{unknown} \times$ depth of unknown solution $= [FeSCN^{2+}]$

$$\times \text{ depth of standard solution} \quad (7)$$

Measure the depths of the matching solutions with a rule and record them. Repeat the measurement for Mixtures 2 through 5, recording the depth of each unknown and that of the standard solution which matches it in intensity.

In this experiment, reactant concentrations are very low. In most localities you can pour the contents of the test tubes down the sink when you have completed your measurements. However, consult your instructor for alternate disposal procedures.

Name —————————————

Section —————————

Data: Determination of the Equilibrium Constant for a Chemical Reaction

Mixture	Volume in mL, 2.00×10^{-3} M Fe(NO₃)₃	Volume in mL, 2.00×10^{-3} KSCN	Volume in mL of Water	Method I Absorbance	Method II Depth in mm		[FeSCN]²⁺
					Standard	Unknown	
1	5.00	1.00	4.00	———	———	———	——— × 10⁻⁴ M
2	5.00	2.00	3.00	———	———	———	——— × 10⁻⁴ M
3	5.00	3.00	2.00	———	———	———	——— × 10⁻⁴ M
4	5.00	4.00	1.00	———	———	———	——— × 10⁻⁴ M
5	5.00	5.00	0.00	———	———	———	——— × 10⁻⁴ M

If Method II was used, $[FeSCN^{2+}]_{standard} = $ ——— $\times 10^{-4}$ M; $[FeSCN^{2+}]$ in Mixtures 1 to 5 found by Equation 7.

Calculations

A. Calculation of K_c assuming the reaction

$$Fe^{3+}(aq) + SCN^-(aq) \rightleftharpoons FeSCN^{2+}(aq) \tag{1}$$

This calculation is most easily done by following Steps 1 through 5 in the discussion. Results are to be entered in the table on the following page.

Step 1. Find the initial number of moles of Fe^{3+} and SCN^- in the mixtures in test tubes 1 through 5. Use Equation 3 and enter the values in the first two columns of the table.

Step 2. Enter the experimentally determined value of $[FeSCN^{2+}]$ at equilibrium for each of the mixtures in the next to last column in the table. Use Equation 3 to find the number of moles of $FeSCN^{2+}$ in each of the mixtures, and enter the values in the fifth column of the table. Note that this is also the number of moles of Fe^{3+} and SCN^- that were used up in the reaction.

(continued on following page)

Step 3. From the number of moles of Fe^{3+} and SCN^- initially present in each mixture, and the number of moles of Fe^{3+} and SCN^- used up in forming $FeSCN^{2+}$, calculate the number of moles of Fe^{3+} and SCN^- that remain in each mixture at equilibrium. Use Equation 4. Enter the results in columns 3 and 4 of the table.

Step 4. Use Equation 3 and the results of Step 3 to find the concentrations of all of the species at equilibrium. The volume of the mixture is 10.00 mL, or 0.0100 liter in all cases. Enter the values in columns 6 and 7 of the table.

Step 5. Calculate K_c for the reaction for each of the mixtures by substituting values for the equilibrium concentrations of Fe^{3+}, SCN^-, and $FeSCN^{2+}$ in Equation 2.

| *Mixture* | *Initial No. Moles* | | *Equilibrium No. Moles* | | | *Equilibrium Concentrations* | | | K_c |
	Fe^{3+}	SCN^-	Fe^{3+}	SCN^-	$FeSCN^{2+}$	$[Fe^{3+}]$	$[SCN^-]$	$[FeSCN^{2+}]$	
1	___	___ $\times 10^{-6}$	___ $\times 10^{-6}$	___ $\times 10^{-6}$	___ $\times 10^{-6}$	___ $\times 10^{-4}$	___ $\times 10^{-4}$	___ $\times 10^{-4}$	___
2	___	___	___	___	___	___	___	___	___
3	___	___	___	___	___	___	___	___	___
4	___	___	___	___	___	___	___	___	___
5	___	___	___	___	___	___	___	___	___

(continued on following page)

(continued)

B. (Optional) In calculating K_c in Part A, we assume, correctly, that the formula of the complex ion is $FeSCN^{2+}$. It is by no means obvious that this is the case and one might have assumed, for instance, that $Fe(SCN)_2^+$ was the species formed. The reaction would then be

$$Fe^{3+}(aq) + 2\ SCN^-(aq) \rightleftharpoons Fe(SCN)_2^+(aq) \tag{5}$$

If we analyze the equilibrium system we have studied, assuming that Reaction 5 occurs rather than Reaction 1, we would presumably obtain nonconstant values of K_c. Using the same kind of procedure as in Part A, calculate K_c for Mixtures 1, 3, and 5 on the basis that $Fe(SCN)_2^+$ is the formula of the complex ion formed by the reaction between Fe^{3+} and SCN^-. As a result of the procedure used for calibrating the system by Method I or Method II, $[Fe(SCN)_2^+]$ will equal *one-half* the $[FeSCN^{2+}]$ obtained for each solution in Part A. Note that *two* moles SCN^- are needed to form *one* mole $Fe(SCN)_2^+$. This changes the expression for K_c. Also, in calculating the equilibrium number of moles SCN^- you will need to subtract ($2 \times$ number of moles $Fe(SCN)_2^+$) from the initial number of moles SCN^-.

Mixture	Initial No. Moles		Equilibrium No. Moles			Equilibrium Concentrations			K_c
	Fe^{3+}	SCN^-	Fe^{3+}	SCN^-	$Fe(SCN)_2^+$	$[Fe^{3+}]$	$[SCN^-]$	$[Fe(SCN)_2^+]$	
1	___	___	___	___	___	___	___	___	___
3	___	___	___	___	___	___	___	___	___
5	___	___	___	___	___	___	___	___	___

On the basis of the results of Part A, what can you conclude about the validity of the equilibrium concept, as exemplified by Equation 2? What do you conclude about the formula of the iron(III) thiocyanate complex ion?

191

Advance Study Assignment: Determination of the Equilibrium Constant
for a Reaction

1. A student mixes 5.00 mL 2.00×10^{-3} M $Fe(NO_3)_3$ with 5.00 mL 2.00×10^{-3} M KSCN. She finds that in the equilibrium mixture the concentration of $FeSCN^{2+}$ is 1.40×10^{-4} M. Find K_c for the reaction $Fe^{3+}(aq) + SCN^-(aq) \rightleftharpoons FeSCN^2(aq)$.

Step 1. Find the number of moles Fe^{3+} and SCN^- initially present. (Use Eq. 3.)

_____ moles Fe^{3+};   _____ moles SCN^-

Step 2. How many moles $FeSCN^{2+}$ are in the mixture at equilibrium? What is the volume of the equilibrium mixture? (Use Eq. 3.)

_____ mL;   _____ moles $FeSCN^{2+}$

How many moles of Fe^{3+} and SCN^- are used up in making the $FeSCN^{2+}$?

_____ moles Fe^{3+};   _____ moles SCN^-

Step 3. How many moles of Fe^{3+} and SCN^- remain in the solution at equilibrium? (Use Eq. 4 and the results of Steps 1 and 2.)

_____ moles Fe^{3+};   _____ moles SCN^-

Step 4. What are the concentrations of Fe^{3+}, SCN^-, and $FeSCN^{2+}$ at equilibrium? What is the volume of the equilibrium mixture? (Use Eq. 3 and the results of Step 3.)

$[Fe^{3+}] =$ _____ M; $[SCN^-] =$ _____ M; $[FeSCN^{2+}] =$ _____ M

_____ mL

Step 5. What is the value of K_c for the reaction? (Use Eq. 2 and the results of Step 4.)

$K_c =$ _____

(continued on following page)

(continued)

2. **(Optional)** Assume that the reaction studied in Problem 1 is $Fe^{3+}(aq) + 2\,SCN^-(aq) \rightleftharpoons Fe(SCN)_2^+(aq)$. Find K_c for this reaction, given the data in Problem 1, except that the equilibrium concentration of $Fe(SCN)_2^+$ is equal to 0.70×10^{-4} M.

   a. Formulate the expression for K_c for the alternate reaction just cited.

   b. Find K_c as you did in Problem 1; take due account of the fact that two moles SCN^- are used up per mole $Fe(SCN)_2^+$ formed.

    Step 1. Results are as in Problem 1.

    Step 2. How many moles of $Fe(SCN)_2^+$ are in the mixture at equilibrium? (You should use Eq. 3.)

          _____ moles $Fe(SCN)_2^+$

      How many moles of Fe^{3+} and SCN^- are used up in making the $Fe(SCN)_2^+$?

          _____ moles Fe^{3+};   _____ moles SCN^-

    Step 3. How many moles of Fe^{3+} and SCN^- remain in solution at equilibrium? Use the results of Steps 1 and 2, noting that no. moles SCN^- at equilibrium = original no. moles $SCN^- - (2 \times$ no. moles $Fe(SCN)_2^+)$.

          _____ moles Fe^{3+};   _____ moles SCN^-

    Step 4. What are the concentrations of Fe^{3+}, SCN^-, and $Fe(SCN)_2^+$ at equilibrium? (Use Eq. 3 and the results of Step 3.)

    $[Fe^{3+}] = $ _____ M; $[SCN^-] = $ _____ M; $[Fe(SCN)_2^+] = $ _____ M

    Step 5. Calculate K_c on the basis that the alternate reaction occurs. (Use the answer to Part 2a.)

          $K_c = $ _____

THE STANDARDIZATION OF A BASIC SOLUTION AND THE DETERMINATION OF THE EQUIVALENT MASS OF AN ACID

When a solution of a strong acid is mixed with a solution of a strong base, a chemical reaction occurs that can be represented by the following net ionic equation:

$$H^+(aq) + OH^-(aq) \rightarrow H_2O$$

This is called a neutralization reaction, and chemists use it extensively to change the acidic or basic properties of solutions. The equilibrium constant for the reaction is about 10^{14} at room temperature, so that the reaction can be considered to proceed completely to the right, using up whichever of the ions is present in the lesser amount and leaving the solution either acidic or basic, depending on whether H^+ or OH^- ion was in excess.

Since the reaction is essentially quantitative, it can be used to determine the concentrations of acidic or basic solutions. A frequently used procedure involves the titration of an acid with a base. In the titration, a basic solution is added from a buret to a measured volume of acid solution until the number of moles of OH^- ion added is just equal to the number of moles of H^+ ion present in the acid. At that point the volume of basic solution that has been added is measured.

Recalling the definition of the molarity, M_A, of species A, we have

$$M_A = \frac{\text{no. moles of A}}{\text{no. liters of solution, } V} \quad \text{or no. moles of A} = M_A \times V \qquad (1)$$

At the end point of the titration,

$$\text{no. moles } H^+ \text{ originally present} = \text{no. moles } OH^- \text{ added} \qquad (2)$$

So, by Equation 1,

$$M_{H^+} \times V_{acid} = M_{OH^-} \times V_{base} \qquad (3)$$

Therefore, if the molarity of either the H^+ or the OH^- ion in its solution is known, the molarity of the other ion can be found from the titration.

The equivalence point or end point in the titration is determined by using a chemical, called an indicator, that changes color at the proper point. The indicators used in acid-base titrations are weak organic acids or bases that change color when they are neutralized. One of the most common indicators is phenolphthalein, which is colorless in acid solutions but becomes red when the pH of the solution becomes 9 or higher.

When a solution of a strong acid is titrated with a solution of a strong base, the pH at the end point will be about 7. At the end point a drop of acid or base added to the solution will change its pH by several pH units, so that phenolphthalein can be used as an indicator in such titrations. If a weak acid is titrated with a strong base, the pH at the equivalence point is somewhat higher than 7, perhaps 8 or 9, and phenolphthalein is still a very satisfactory indicator. If, however, a solution of a weak base such as ammonia is titrated with a strong acid, the pH will be a unit or two less than 7 at the end point, and phenolphthalein will not be as good an indicator for that titration as, for example, methyl red, whose color changes from red to yellow as the pH changes from about 4 to 6. Ordinarily, indicators will be chosen so that their color change occurs at about the pH at the equivalence point of a given acid-base titration.

In this experiment you will determine the molarity of OH^- ion in an NaOH solution by titrating that solution against a standardized solution of HCl. Since in these solutions one mole of acid in solution furnishes one mole of H^+ ion and one mole of base produces one mole of OH^- ion, $M_{HCl} = M_{H^+}$ in the acid solution, and $M_{NaOH} = M_{OH^-}$ in the basic solution. Therefore, the titration will allow you to find M_{NaOH} as well as M_{OH^-}.

In the second part of this experiment you will use your standardized NaOH solution to titrate a sample of a pure solid organic acid. By titrating a weighed sample of the unknown acid with your standardized NaOH solution you can, by Equation 2, find the number of moles of H^+ ion that your sample can furnish. From the mass of your sample and the number of moles of H^+ it contains, you can calculate the number of grams of solid acid, EM, that would contain one mole of H^+ ion.

$$EM = \frac{\text{no. grams acid}}{\text{no. moles } H^+ \text{ furnished}} \tag{4}$$

The value of EM for an acid is called the equivalent mass of the acid and is equal to the number of grams of acid per mole of available H^+ ion:

$$\text{one equivalent mass acid} \rightarrow \text{one mole } H^+ \text{ ion} \tag{5}$$

The equivalent mass, EM, of an acid may or may not equal the molar mass of the acid, MM. The reason is almost, but not quite, obvious. Let us consider three acids, with the molecular formulas HX, H_2Y, and H_3Z; presuming that in the titration all of the H^+ ion in the acid is neutralized:

one mole HX $\rightarrow$ one mole H^+ Therefore, by Equation 5, MM = EM

one mole H_2Y $\rightarrow$ two moles H^+ "   MM = 2 × EM

one mole H_3Z $\rightarrow$ three moles H^+ "   MM = 3 × EM

In all cases the molar mass and the equivalent mass are related by simple equations, but in order to find the molar mass from the equivalent mass we need to know the molecular formula of the acid.

Experimental Procedure

Note: This experiment is relatively long unless you know precisely what to do. Study the experiment carefully before coming to class, so that you don't have to spend a lot of time finding out what the experiment is all about.

Obtain two burets and a sample of solid unknown acid from the stockroom.

A. Standardization of NaOH Solution

Into a small graduated cylinder draw about 7 mL of the stock 6 M NaOH solution provided in the laboratory and dilute to about 400 mL with distilled water in a 500-mL Florence flask. Stopper the flask tightly and mix the solution thoroughly at intervals over a period of at least 15 minutes before using the solution.

Draw into a clean, *dry* 125-mL Erlenmeyer flask about 75 mL of standardized HCl solution (about 0.1 M) from the stock solution on the reagent shelf. This amount should provide all the standard acid you will need; do not waste it. Record the molarity of the HCl.

Prepare for the titration by using the procedure described in Experiment 6 and Appendix IV. The purpose of this procedure is to make sure that the solution in each buret has the same molarity as it has in the container from which it was poured. Clean the two burets and rinse them with distilled water. Then rinse one buret three times with a few milliliters of the HCl solution, in each case thoroughly wetting the walls of the buret with the solution and then letting it out through the stopcock. Fill the buret with HCl; open the stopcock momentarily to fill the tip. Proceed to clean and fill the other buret with your NaOH solution in a similar manner. Put the acid buret, A, on the left side of your buret clamp, and the base buret, B, on the right side. Check to see that your burets do not leak and that there are no air bubbles in either buret tip. Read and record the levels in the two burets to 0.02 mL.

Draw about 25 mL of the HCl solution from the buret into a clean 250-mL Erlenmeyer flask; add to the flask about 25 mL of distilled H_2O and two or three drops of phenolphthalein indicator solution. Place a white sheet of paper under the flask to aid in the detection of any color change. Add the NaOH solution intermittently from its buret to the solution in the flask, noting the pink phenolphthalein color that appears and disappears as the drops hit the liquid and are mixed with it. Swirl the liquid in the flask gently and continuously as you add the NaOH solution. When the pink color begins to persist, slow down the rate of addition of NaOH. In the final stages of the titration add the NaOH drop by drop until the entire solution just turns a pale pink color that will persist for about 30 seconds. If you go past the end point and obtain a red solution, add a few drops of the HCl solution to remove the color, and then add NaOH a drop at a time until the pink color persists. Carefully record the *final* readings on the HCl and NaOH burets.

To the 250-mL Erlenmeyer flask containing the titrated solution, add about 10 mL more of the standard HCl solution. Titrate this as before with the NaOH to an end point, and carefully record both buret readings once again. To this solution add about 10 mL more HCl and titrate a third time with NaOH.

You have now completed three titrations, with *total* HCl volumes of about 25, 35, and 45 mL. Using Equation 3, calculate the molarity of your base, M_{OH^-} for each of the three titrations. In each case, use the *total volumes* of acid and base that were added up to that point. At least two of these molarities should agree to within 1%. If they do, proceed to the next part of the experiment. If they do not, repeat these titrations until two calculated molarities do agree.

B. Determination of the Equivalent Mass of an Acid

Weigh the vial containing your solid acid on the analytical balance to ± 0.0001 g. Carefully pour out about half the sample into a clean but not necessarily dry 250-mL Erlenmeyer flask. Weigh the vial accurately. Add about 50 mL of distilled water and two or three drops of phenolphthalein to the flask. The acid may be relatively insoluble, so don't worry if it doesn't all dissolve.

Fill your NaOH buret with the (now standardized) NaOH solution. Add the standard HCl to your HCl buret until it is about half full. Read both levels carefully and record them.

Titrate the solution of the solid acid with NaOH. As the acid is neutralized it will tend to dissolve in the solution. If the acid appears to be relatively insoluble, add NaOH until the pink color persists, and then swirl to dissolve the solid. If the solid still will not dissolve, and the solution remains pink, add 25 mL of ethanol to increase the solubility. If you go past the end point, add HCl as necessary. The final pink end point should appear on addition of one drop of NaOH. Record the final levels in the NaOH and HCl burets.

Pour the rest of your acid sample into a clean 250-mL Erlenmeyer flask, and weigh the vial accurately. Titrate this sample of acid as before with the NaOH and HCl solutions.

If you use HCl in these titrations, and you probably will, the calculations needed are a bit more complicated than in the standardization of the NaOH solution. To find the number of moles of H^+ ion in the solid acid, you must subtract the number of moles of HCl used from the number of moles of NaOH. For a back-titration, which is what we have in this case:

no. moles H^+ in solid acid = no. moles OH^- in NaOH soln.

$$- \text{ no. moles } H^+ \text{ in HCl soln.} \quad (6)$$

For volumes in milliliters, this equation takes the form:

$$\text{no. moles } H^+ \text{ in solid acid} = \frac{M_{NaOH} \times V_{NaOH}}{1000} - \frac{M_{HCl} \times V_{HCl}}{1000} \quad (7)$$

C. Determination of K_a for the Unknown Acid (Optional)

Your instructor will tell you in advance if you are to do this part of the experiment. If you do this part, you will need to **save** one of the titrated solutions from Part B. Given that titrated solution, it is easy to prepare one in which $[H^+]$ ion is equal to K_a for your acid.

A weak acid, HX, dissociates in water solution according to the equation:

$$HX(aq) \rightleftarrows H^+(aq) + X^-(aq) \quad (8)$$

The equilibrium constant for this reaction is called the acid dissociation constant for HX and is given the symbol K_a. A solution of HX will obey the equilibrium condition given by the equation:

$$K_a = \frac{[H^+][X^-]}{[HX]} \quad (9)$$

Your acid is a weak acid, so will follow both of the above equations.

In the titration in Part B you converted a solution of HX into one containing NaX by adding NaOH to the end point. If, to your titrated solution, you now add some HCl, it will convert some of the X^- ions in the NaX solution back to HX. If the number of moles of HCl you add equals *one-half* the number of moles of H^+ in your original sample, then *half* of the X^- ions will be converted to HX. In the resulting solution, $[HX]$ will equal $[X^-]$, and so, by Equation 9,

$$K_a = [H^+] \quad \text{in the half-neutralized solution} \quad (10)$$

Using the approach we have outlined, use your titrated solution to prepare one in which $[HX]$ equals $[X^-]$. Measure the pH of that solution with a pH meter, using the procedure described in Appendix IV. From that value, calculate K_a for your unknown acid.

Name _____   *Section* _____

Data: Standardization of a Basic Solution
·Determination of the EM of an Acid

A. Standardization of NaOH Solution

	Trial 1	*Trial 2*	*Trial 3*
Initial reading, HCl buret	_____ mL		
Final reading, HCl buret	_____ mL	_____ mL	_____ mL
Initial reading, NaOH buret	_____ mL		
Final reading, NaOH buret	_____ mL	_____ mL	_____ mL

B. Equivalent Mass of Unknown Acid

Mass of vial plus contents   _____ g

Mass of vial plus contents less Sample 1   _____ g

Mass less Sample 2   _____ g

	Trial 1	*Trial 2*
Initial reading, NaOH buret	_____ mL	_____ mL
Final reading, NaOH buret	_____ mL	_____ mL
Initial reading, HCl buret	_____ mL	_____ mL
Final reading, HCl buret	_____ mL	_____ mL

Calculations

A. Standardization of NaOH Solution

	Trial 1	*Trial 2*	*Trial 3*
Total volume HCl	_____ mL	_____ mL	_____ mL
Total volume NaOH	_____ mL	_____ mL	_____ mL

(continued on following page)

(continued)

Molarity, M_{HCl}, of standardized HCl   _____ M

Molarity, M_{H^+}, in standardized HCl   _____ M

By Equation 3,

$$M_{H^+} \times V_{acid} = M_{OH^-} \times V_{base} \quad \text{or} \quad M_{OH^-} = M_{H^+} \times \frac{V_{HCl}}{V_{NaOH}} \tag{3}$$

Use Equation 3 to find the molarity, M_{OH^-}, of the NaOH solution. Note that the volumes do not need to be converted to liters, since we use the volume ratio.

       Trial 1   *Trial 2*   *Trial 3*

M_{OH^-} _____ M   _____ M   _____ M   (should agree with 1%)

The molarity of the NaOH will equal M_{OH^-}, since one mole NaOH $\rightarrow$ one mole OH^-.

Average molarity of NaOH solution, M_{NaOH} _____ M

B. Equivalent Mass of Unknown Acid

	Trial 1	Trial 2
Mass of sample	_____ g	_____ g
Volume NaOH used	_____ mL	_____ mL
No. moles NaOH $= \dfrac{V_{NaOH} \times M_{NaOH}}{1000}$	_____	_____
Volume HCl used	_____ mL	_____ mL
No. moles HCl $= \dfrac{V_{HCl} \times M_{HCl}}{1000}$	_____	_____
No. moles H^+ in sample (by Eq. 6)	_____	_____
EM $= \dfrac{\text{no. grams acid}}{\text{no. moles } H^+}$	_____ g	_____ g
Unknown no.	_____	

C. Determination of K_a for the Unknown Acid (Optional)

No. moles H^+ in sample (from Part B)   _____

No. moles HCl to be added _____   Volume of HCl added _____ mL

pH of half-neutralized solution _____ K_a of acid _____

Advance Study Assignment: Equivalent Mass of an Unknown Acid

1. 7.0 mL of 6.0 M NaOH are diluted with water to a volume of 400 mL. You are asked to find the molarity of the resulting solution.

   a. First find out how many moles of NaOH there are in 7.0 mL of 6.0 M NaOH. Use Equation 1. Note that the volume must be in liters.

   _____ moles

   b. Since the total number of moles of NaOH is not changed on dilution, the molarity after dilution can also be found by Equation 1, using the final volume of the solution. Calculate that molarity.

   _____ M

2. In an acid-base titration, 24.88 mL of an NaOH solution are needed to neutralize 26.43 mL of a 0.1049 M HCl solution. To find the molarity of the NaOH solution, we can use the following procedure:

   a. First note the value of M_{H^+} in the HCl solution.

   _____ M

   b. Find M_{OH^-} in the NaOH solution. (Use Eq. 3.)

   _____ M

   c. Obtain M_{NaOH} from M_{OH^-}.

   _____ M

3. A 0.2349 g sample of an unknown acid requires 33.66 mL of 0.1086 M NaOH for neutralization to a phenolphthalein end point. There are 0.42 mL of 0.1049 M HCl used for back-titration.

   a. How many moles of OH^- are used? How many moles of H^+ from HCl?

   _____ moles OH^- _____ moles H^+

   b. How many moles of H^+ are there in the solid acid? (Use Eq. 6.)

   _____ moles H^+ in solid

   c. What is the equivalent mass of the unknown acid? (Use Eq. 4.)

   _____ g

pH MEASUREMENTS—BUFFERS AND THEIR PROPERTIES

One of the more important properties of an aqueous solution is its concentration of hydrogen ion. The H^+ or H_3O^+ ion has great effect on the solubility of many inorganic and organic species, on the nature of complex metallic cations found in solutions, and on the rates of many chemical reactions. It is important that we know how to measure the concentration of hydrogen ion and understand its effect on solution properties.

For convenience the concentration of H^+ ion is frequently expressed as the pH of the solution rather than as molarity. The pH of a solution is defined by the following equation:

$$pH = -\log[H^+] \tag{1}$$

where the logarithm is taken to the base 10. If $[H^+]$ is 1×10^{-4} moles per liter, the pH of the solution is, by the equation, 4. If the $[H^+]$ is 5×10^{-2} M, the pH is 1.3.

Basic solutions can also be described in terms of pH. In water solutions the following equilibrium relation will always be obeyed:

$$[H^+] \times [OH^-] = K_w = 1 \times 10^{-14} \text{ at } 25°C \tag{2}$$

In distilled water $[H^+]$ equals $[OH^-]$, so, by Equation 2, $[H^+]$ must be 1×10^{-7} M. Therefore, the pH of distilled water is 7. Solutions in which $[H^+] > [OH^-]$ are said to be acidic and will have a pH < 7; if $[H^+] < [OH^-]$, the solution is basic and its pH > 7. A solution with a pH of 10 will have a $[H^+]$ of 1×10^{-10} M and a $[OH^-]$ of 1×10^{-4} M.

We measure the pH of a solution experimentally in two ways. In the first of these we use a chemical called an indicator, which is sensitive to pH. These substances have colors that change over a relatively short pH range (about two pH units) and can, when properly chosen, be used to determine roughly the pH of a solution. Two very common indicators are litmus, usually used on paper, and phenolphthalein, the most common indicator in acid-base titrations. Litmus changes from red to blue as the pH of a solution goes from about 6 to about 8. Phenolphthalein changes from colorless to red as the pH goes from 8 to 10. A given indicator is useful for determining pH only in the region in which it changes color. Indicators are available for measurement of pH in all the important ranges of acidity and basicity. By matching the color of a suitable indicator in a solution of known pH with that in an unknown solution, one can determine the pH of the unknown to within about 0.3 pH units.

The other method for finding pH is with a device called a pH meter. In this device two electrodes, one of which is sensitive to $[H^+]$, are immersed in a solution. The potential between the two electrodes is related to the pH. The pH meter is designed so that the scale will directly furnish the pH of the solution. A pH meter gives much more precise measurement of pH than does a typical indicator and is ordinarily used when an accurate determination of pH is needed.

Some acids and bases undergo substantial ionization in water, and are called strong because of their essentially complete ionization in reasonably dilute solutions. Other acids and bases, because of incomplete ionization (often only about 1% in 0.1 M solution), are called weak. Hydrochloric acid, HCl, and sodium hydroxide, NaOH, are typical examples of a strong acid and a strong base. Acetic acid, $HC_2H_3O_2$, and ammonia, NH_3, are classic examples of a weak acid and a weak base.

A weak acid will ionize according to the Law of Chemical Equilibrium:

$$HB(aq) \rightleftharpoons H^+(aq) + B^-(aq) \tag{3}$$

At equilibrium,

$$\frac{[H^+][B^-]}{[HB]} = K_a \tag{4}$$

K_a is a constant characteristic of the acid HB; in solutions containing HB, the product of concentrations in the equation will remain constant at equilibrium independent of the manner in which the solution was made. A similar relation can be written for solutions of a weak base.

The value of the ionization constant K_a for a weak acid can be found experimentally in several ways. In Experiment 22 we used a very simple method for finding K_a. In general, however, we need to find the concentrations of each of the species in Equation 4 by one means or another. In this experiment we will determine K_a for a weak acid in connection with our study of the properties of those solutions we call buffers.

Salts that can be formed by the reaction of strong acids and bases—such as NaCl, KBr, or $NaNO_3$—ionize completely but do not react with water when in solution. They form neutral solutions with a pH of about 7. When dissolved in water, salts of *weak* acids or *weak* bases furnish ions that tend to react to some extent with water, producing molecules of the weak acid or base and liberating some OH^- or H^+ ion to the solution.

If HB is a weak acid, the B^- ion produced when NaB is dissolved in water will react with water to some extent, according to the equation

$$B^-(aq) + H_2O \rightleftharpoons HB(aq) + OH^-(aq) \tag{5}$$

Solutions of sodium acetate, $NaC_2H_3O_2$, the salt formed by reaction of sodium hydroxide with acetic acid, will be slightly *basic* because of the reaction of $C_2H_3O_2^-$ ion with water to produce $HC_2H_3O_2$ and OH^-. Because of the analogous reaction of the NH_4^+ ion with water to form H_3O^+ ion, solutions of ammonium chloride, NH_4Cl, will be slightly *acidic*.

Salts of most transition metal ions are acidic. A solution of $CuSO_4$ or $FeCl_3$ will typically have a pH equal to 5 or lower. The salts are completely ionized in solution. The acidity comes from the fact that the cation is hydrated (e.g., $Cu(H_2O)_4^{2+}$ or $Fe(H_2O)_6^{3+}$). The large $+$ charge on the metal cation attracts electrons from the O—H bonds in water, weakening them and producing some H^+ ions in solution; with $CuSO_4$ solutions the reaction would be

$$Cu(H_2O)_4^{2+}(aq) \rightleftharpoons Cu(H_2O)_3OH^+(aq) + H^+(aq) \tag{6}$$

BUFFERS

Some solutions, called buffers, are remarkably resistant to pH changes. Water is not a buffer, since its pH is very sensitive to addition of any acidic or basic species. Even bubbling your breath through a straw into distilled water can change its pH by at least 1 unit, just due to the small amount of CO_2 in exhaled air. With a good buffer solution, you could blow your exhaled air into it for half an hour and not change the pH appreciably. All living systems contain buffer solutions, since stability of pH is essential for the occurrence of many of the biochemical reactions that go on to maintain the living organism.

There is nothing mysterious about what one needs to make a buffer. All that is required is a solution containing a weak acid and its conjugate base. An example of such a solution is one containing the weak acid HB, and the B^- ion, its conjugate base, obtained by dissolving the salt NaB in water.

The pH of such a buffer is established by the relative concentrations of HB and B^- in the solution. In such a solution $[H^+]$ can be calculated by manipulating Equation 4:

$$[H^+] = K_a \times \frac{[HB]}{[B^-]} \tag{4a}$$

If, for example, we mix, 500 mL of 0.10 M HB with 500 mL 0.10 M NaB, we will have a typical buffer, containing an acid and its conjugate base. The $[H^+]$ in this solution is easily found. Since we have equal amounts of HB and B^- present, their concentrations are equal, and by Equation 4a, $[H^+]$ equals K_a.

You might wonder why $[HB]$ and $[B^-]$ do not change when the species are mixed. Actually, they do, very very slightly, just enough to generate enough H^+ ion to satisfy the condition imposed by Equation 4a. Ordinarily, however, K_a is small, so $[H^+]$ is also small. If $K_a = 1 \times 10^{-5}$, $[H^+]$ will be 1×10^{-5} M, and so, in our example, where we have 1 L of solution, we will have 1×10^{-5} moles of H^+. This means that 1×10^{-5} moles of HB dissociate, out of 0.050 moles initially present, so only a negligible decrease in $[HB]$ occurs, and only a tiny increase in $[B^-]$ as a result of the reaction to form the equilibrium system.

From the above discussion, we can conclude that the acid and conjugate base in a buffer *do not* react appreciably when mixed, so the relative concentrations can be calculated from the way the buffer was put together.

Another interesting property of a buffer is that its pH does not change appreciably on dilution. If we look again at the buffer in our example, if we increased the volume from 1 L to 5 L by adding water, the ratio of $[HB]$ to $[B^-]$ would not change, and since that ratio fixes $[H^+]$, the pH would not change.

We can adjust the pH of the buffer, within limits, to bring it to some desired value. In our example, if K_a for the buffer is 1.0×10^{-5}, the pH of the buffer solution would be 5.0. If we wish to make a buffer of pH equal to 4.5, we need to simply select volumes of the acid and conjugate base such that the resultant ratio of $[HB]$ to $[B^-]$ would make $[H^+]$ equal to $10^{-4.5}$, or 3.2×10^{-5} M. Then, by Equation 4a,

$$3.2 \times 10^{-5} = 1.0 \times 10^{-5} \times \frac{[HB]}{[B^-]} \quad \text{and} \quad \frac{[HB]}{[B^-]} = 3.2$$

So, to make the desired buffer we could use 320 mL 0.10 M HB and 100 mL 0.10 M NaB. Or, if our stock solutions were of different concentrations, we would select reagent volumes such that the number of *moles* of HB used would be 3.2 times as large as the number of *moles* of B^-.

The reason that a buffer has a stable pH is that its two components can "soak up" added H^+ or OH^- ions. If we add a little HCl, a strong acid, to our buffer, the following reaction will occur:

$$H^+(aq) + B^- \rightarrow HB(aq) + H_2O \tag{7}$$

If we added a little NaOH, a strong base, it will react with the HB present:

$$OH^-(aq) + HB(aq) \rightarrow B^-(aq) + H_2O \tag{8}$$

As a result of these reactions, $[HB]$ and $[B^-]$ will change slightly, but if the amounts of H^+ and OH^- ions that are added are *small* as compared to the amounts of HB and B^- present in the buffer, the effect on the pH will be small since the ratio of $[HB]$ to $[B^-]$ will not change much.

The range over which a buffer is useful is limited to about 2 pH units. In the example we used earlier, if we mixed 500 mL 0.10 M HB with 50 mL 0.10 M NaB, in the buffer $[HB]/[B^-]$ would be 10, and so $[H^+]$ would be 1×10^{-4} M and the pH would be 4.0. This buffer could deal with added NaOH much better than with added HCl, since the amount of available HB is much greater than that of B^-. Indeed, if we add enough HCl to react with all of the B^- present, the buffer would be "exhausted," since it would contain only HB, and any excess HCl would produce a pH with just about the same value as if the HCl were added to water. Similar behavior would occur if we made the buffer in such a way that $[HB]/[B^-]$ were equal to 0.1. Then the pH would be 6, and the buffer would have very little capacity for added NaOH.

In the first part of this experiment you will determine the approximate pH of several solutions by using acid-base indicators. Then you will find the pH of some other solutions with a pH meter. The rest of the experiment will deal with the properties of buffer solutions. You will be working with one acid-conjugate base buffer system. You will note the effect on pH of changing the composition of the buffer, and use the data obtained to find the K_a of the acid. The stability of the pH as we add small amounts of acid and base will be examined. The effect of dilution on the pH will also be noted. We will then exhaust a buffer by adding an excess amount of acid or base. Finally, we will prepare one or two buffers having specific pH values.

Experimental Procedure

You may work in pairs on the first three parts of this experiment.

A. Determination of pH by the Use of Acid-Base Indicators

To each of five small test tubes add about 1 mL 0.1 M HCl (about 1/2-inch depth in tube). To each tube add a drop or two of one of the indicators in Table 23.1, one indicator to a tube. Note the color of the solution you obtain in each case. By comparing the colors you observe with the information in Table 23.1, estimate the pH of the solution to within a range of one pH unit, say 1 to 2, or 4 to 5. In making your estimate, note that the color of an indicator is most indicative of pH in the region where the indicator is changing color.

Repeat the procedure with each of the following solutions:

$$0.1 \text{ M NaH}_2\text{PO}_4 \qquad 0.1 \text{ M HC}_2\text{H}_3\text{O}_2 \qquad 0.1 \text{ M ZnSO}_4$$

Record the colors you observe and the pH range for each solution.

Table 23.1

| | *Useful pH Range (Approximate)* | | | | | | | |
Indicator	0	1	2	3	4	5	6	7
Methyl violet	yellow [ ] violet							
Thymol blue	red [ ] yellow							
Methyl yellow	red [ ] yellow							
Congo red	violet [ ] orange-red							
Bromcresol green	yellow [ ] blue							

B. Measurement of the pH of Some Typical Solutions

In the rest of this experiment we will use pH meters to find pH. Your instructor will show you how to operate your meter. The electrodes may be fragile, so use due caution when handling the electrode probe. See Appendix IV for a discussion of pH meters.

Using a 25-mL sample in a 150-mL beaker, measure and record the pH of a 0.1 M solution of each of the following substances:

$$NaCl \qquad Na_2CO_3 \qquad NaC_2H_3O_2 \qquad NaHSO_4$$

Rinse the electrode probe in distilled water between measurements. After you have completed a measurement, add a drop or two of bromcresol green to the solution and record the color you obtain.

Some of the solutions are nearly neutral, others acidic or basic. For each solution having a pH less than 6 or greater than 8, write a net ionic equation that explains qualitatively why the observed pH value is reasonable.

Then write a rationale for the colors obtained with bromcresol green with these solutions.

C. Some Properties of Buffers

On the lab bench we have 0.10 M stock solutions that can be used to make three different common buffer systems. These are

$HC_2H_3O_2 - C_2H_3O_2^-$	$NH_4^+ - NH_3$	$HCO_3^- - CO_3^{2-}$
acetic acid–acetate ion	ammonium ion–ammonia	hydrogen carbonate–carbonate

The sources of the ions will be sodium and ammonium salts containing those ions. Select **one** of these buffer systems for your experiment.

1. Using a graduated cylinder, measure out 15 mL of the acid component of your buffer into a 100-mL beaker. The acid will be in one of the following solutions: 0.10 M $HC_2H_3O_2$, 0.10 M NH_4Cl, or 0.10 M $NaHCO_3$. Rinse out the graduated cylinder with distilled water and use it to add 15 mL of the conjugate base of your buffer. Measure the pH of your mixture and record it on the Data page. Calculate K_a for the acid.
2. Add 30 mL water to your buffer mixture, mix, and pour half of the resulting solution into another 100-mL beaker. Measure the pH of the diluted buffer. Calculate K_a once again. Add five drops of 0.10 M NaOH to the diluted buffer and measure the pH again. To the other half of the diluted buffer add 5 drops 0.10 M HCl, and again measure the pH. Record your results.
3. Make a buffer mixture containing 2 mL of the acid component and 20 mL of the solution containing the conjugate base. Mix, and measure the pH. Calculate a third value for K_a. To that solution add 3 mL 0.10 M NaOH, which should exhaust the buffer. Measure the pH.
4. Put 25 mL distilled water into a 100-mL beaker. Measure the pH. Add 5 drops 0.10 M HCl and measure the pH again. To that solution add 10 drops 0.10 M NaOH, mix, and measure the pH.
5. Select a pH different from any of those you observed in your experiments. Design a buffer which should have that pH by selecting appropriate volumes of your acidic and basic components. Make up the buffer and measure its pH.

D. Preparing a Buffer from an Unknown Acid Solution (Optional)

So far in this experiment we have made buffers by mixing solutions of a weak acid and its conjugate base. It is possible to prepare buffers by adding a strong base to solutions containing a weak acid. Reaction 8 will occur, quantitatively, so we will produce the same number of moles of B^- as we add of NaOH. If we add *half* as many moles of OH^- as we have of HB, the final solution will be half-neutralized, and will be a buffer in which [HB] equals [B^-]. This solution will be completely equivalent to the one we used in our example in which we mixed equal amounts of HB and NaB.

In this part of the experiment we will furnish you with an unknown containing a 0.50 M solution of a weak acid. Using that solution and some 0.10 M NaOH, you will be asked to design and prepare a buffer with a particular pH. The following procedure is suggested:

1. Dilute your unknown to 0.10 M by adding 10 mL of your sample to 40 mL distilled water and mixing thoroughly.
2. Mix 20 mL of your unknown acid with 10 mL of 0.10 M NaOH. Measure the pH of the resulting half-neutralized buffer. Calculate K_a for your unknown acid.
3. Given the pH of the buffer you need to design, and the value of K_a you just found, calculate the value of $\dfrac{[HB]}{[B^-]}$ that is needed in the buffer.

Noting that $\dfrac{[HB]}{[B^-]} = \dfrac{\text{no. moles HB}}{\text{no. moles B}^-}$ in the buffer, find the volumes of 0.10 M NaOH and 0.10 M HB that will produce the required ratio. This is perhaps most easily done by arbitrarily deciding to add 10 mL of the NaOH to a volume of the HB solution. The number of moles of B^- produced by Reaction 8 will equal the number of moles of OH^- in the 10 mL of NaOH, and will also equal the number of moles of HB that will be used up in the reaction with OH^-.

The volume of 0.10 M HB you select must contain the number of moles of HB present in the final buffer *plus* the number of moles used up in producing the B^- that is in the buffer. Knowing the *total* number of moles of HB you need to make the buffer, calculate the volume of the 0.10 M HB that is required. Mix that volume with 10 mL of the 0.10 M NaOH and measure the pH.

Observations, Calculations, and Explanations: pH: Buffers and Their Properties

A. Determination of pH Using Acid-Base Indicators

| *Indicator* | *Color with 0.1 M Solution of* | | | |
	HCl	*NaH$_2$PO$_4$*	*HC$_2$H$_3$O$_2$*	*ZnSO$_4$*
Methyl violet	——————	——————	——————	——————
Thymol blue	——————	——————	——————	——————
Methyl yellow	——————	——————	——————	——————
Congo red	——————	——————	——————	——————
Bromcresol green	——————	——————	——————	——————
pH range	——————	——————	——————	——————

Circle the observation(s) for each solution that was most useful in estimating the pH range.

B. Measurement of the pH of Some Typical Solutions

Record the pH and the color observed with bromcresol green for each of the 0.1 M solutions that were tested.

	NaCl	*Na$_2$CO$_3$*	*NaC$_2$H$_3$O$_2$*	*NaHSO$_4$*
pH	——————	——————	——————	——————
Color	——————	——————	——————	——————

For any two solutions having a pH less than 6 or greater than 8, write a net ionic equation to explain qualitatively why the solution has that pH.

Solution —————   Equation ——————————————————————————————

Solution —————   Equation ——————————————————————————————

Explain why the color observed with bromcresol green for each of the four solutions is reasonable, given the pH.

(continued on following page)

(continued)

C. Some Properties of Buffers

Buffer system selected _____ HB is _____ (name the acid)

1. pH of buffer _____ [H$^+$] _____ M K$_a$ (by Eq. 4a) _____

2. pH of diluted buffer _____ [H$^+$] _____ M   K$_a$ _____

    pH after addition of 5 drops of NaOH _____

    pH after addition of 5 drops of HCl _____

3. pH of buffer in which $\dfrac{[HB]}{[B^-]} = 0.10$ _____   K$_a$ _____

    pH after addition of excess NaOH _____

4. pH of distilled water _____

    pH after addition of 5 drops HCl _____

    pH after addition of 10 drops NaOH _____

5. pH of buffer solution to be prepared _____

Average value of K$_a$ (as calculated in Parts 1, 2, and 3) _____

$\dfrac{[HB]}{[B^-]}$ in buffer (from Eq. 4a) _____

$\dfrac{\text{Volume 0.10 M HB}}{\text{Volume 0.10 M NaB}}$ needed in buffer _____

Volume 0.10 M HB used _____ mL   Volume 0.10 M NaB used _____ mL

pH of prepared buffer _____

D. Preparing a Buffer from an Unknown Acid Solution (Optional)

pH of buffer to be designed and prepared _____

pH of half-neutralized solution ([HB] = [B$^-$]) _____

[H$^+$] in that solution _____ M   K$_a$ of unknown acid _____

$$\dfrac{[HB]}{[B^-]} \text{ needed in buffer} = \underline{\quad\quad} = \dfrac{\text{no. moles HB in buffer}}{\text{no. moles B}^- \text{ in buffer}} \tag{9}$$

Volume of 0.10 M NaOH to be added to the acid solution _____10.0_____ mL

(continued)

No. moles OH$^-$ in that volume   _____ moles

No. moles B$^-$ produced and present in final buffer   _____ moles

No. moles HB that react with the added NaOH   _____ moles

No. moles HB that must be present in final buffer (Eq. 9)   _____ moles

Total number of moles of HB needed to make up the buffer   _____ moles

Volume of 0.10 M HB required   _____ mL

pH of prepared buffer   _____

Unknown No.   _____

Advance Study Assignment: pH Measurements and the Properties of Buffers

1. A solution of a weak acid was tested with the indicators used in this experiment. The colors observed were as follows:

Methyl violet	violet	Congo red	violet
Thymol blue	yellow	Bromcresol green	yellow
Methyl yellow	orange		

    What is the approximate pH of the solution?

    _____

2. The pH of a 0.10 M HCN solution is 5.2.

    a. What is $[H^+]$ in that solution?

    _____ M

    b. What is $[CN^-]$? What is $[HCN]$? (Where do the H^+ and CN^- ions come from?)

    _____ M

    c. What is the value of K_a for HCN? (Eq. 4.)

    _____

3. Formic acid, HFor, has a K_a value equal to about 1.8×10^{-4}. A student is asked to prepare a buffer having a pH of 3.40 from a 0.10 M HFor and a 0.10 M NaFor solution. How many milliliters of the NaFor solution should she add to 20 mL of the 0.10 M HFor to make the buffer?

    _____ mL

4. When 5 drops of 0.10 M NaOH were added to 20 mL of the buffer in problem 3, the pH went from 3.40 to 3.43. Write a net ionic equation to explain why the pH didn't go up to about 10, as it would have if that amount of NaOH were added to distilled water or to 20 mL 0.00040 M HCl, which also would have a pH of 3.40.

xperiment 24

DETERMINATION OF THE SOLUBILITY PRODUCT OF PbI$_2$

In this experiment you will determine the solubility product of lead iodide, PbI$_2$. Lead iodide is relatively insoluble, having a solubility of less than 0.002 mole per liter at 20°C. The equation for the solution reaction of PbI$_2$ is

$$PbI_2(s) \rightleftharpoons Pb^{2+}(aq) + 2I^-(aq) \qquad (1)$$

The solubility product expression associated with this reaction is

$$K_{sp} = [Pb^{2+}][I^-]^2 \qquad (2)$$

Equation 2 implies that in any system containing solid PbI$_2$ in equilibrium with its ions, the product of [Pb^{2+}] times [I$^-$]^2 will at a given temperature have a fixed magnitude, independent of how the equilibrium system was initially made up.

The equilibrium system can be established in many different ways. Perhaps the most obvious is to simply dissolve PbI$_2$ in water. Reaction 1 will proceed to the right until the system reaches equilibrium. The concentrations of Pb^{2+} and I$^-$ ions will be related by the stoichiometry of Reaction 1, so if you can find the concentration of either ion, the other follows; from the two concentrations you can calculate K_{sp} for PbI$_2$ by substituting into Equation 2. In the last part of this experiment we will find K_{sp} by this method.

It is also possible to set up the equilibrium system by mixing two solutions, one containing Pb(NO$_3$)$_2$ and the other containing KI. On mixing these solutions, the Pb^{2+} and I$^-$ ions in those solutions react to form PbI$_2$, and Reaction 1 proceeds to the left. Precipitation of PbI$_2$ will occur until equilibrium is reached. At that point the condition in Equation 2 will be satisfied. In the first part of this experiment, known volumes of standard solutions of Pb(NO$_3$)$_2$ and KI will be mixed in several different proportions. The yellow precipitate of PbI$_2$ formed will be allowed to come to equilibrium with the solution. We will then measure the value of [I$^-$] experimentally. The [Pb^{2+}] will be calculated from the initial composition of the system, the measured value of [I$^-$], and the stoichiometric relationship between Pb^{2+} and I$^-$ that exists as Reaction 1 proceeds to the left. We again find K_{sp} for PbI$_2$ by Equation 2.

The concentration of I$^-$ ion will be found spectrophotometrically, as in Experiment 21. Although the iodide ion is not colored, it is relatively easily oxidized to I$_2$, which is brown in water solution. Our procedure will be to separate the solid PbI$_2$ from the solution and then to oxidize the I$^-$ in solution with potassium nitrite, KNO$_2$, under slightly acidic conditions, where the conversion to I$_2$ is quantitative. Although the concentration of I$_2$ will be rather low in the solutions you will prepare, the absorption of light by I$_2$ in the vicinity of 525 nm is sufficiently intense to make accurate analyses possible.

In all of the solutions prepared, potassium nitrate KNO$_3$ (note the distinction between KNO$_2$ and KNO$_3$) will be present as an inert salt. This salt serves to keep the ionic strength of the solution essentially constant at 0.2 M and promotes the formation of well-defined crystalline precipitates of PbI$_2$.

215

Experimental Procedure

From the stock solutions that are available, measure out about 35 mL of 0.0120 M Pb(NO$_3$)$_2$ in 0.20 M KNO$_3$ into a small beaker. To a second small beaker add 30 mL 0.0300 M KI in 0.20 M KNO$_3$ and, to a third, add 10 mL 0.20 M KNO$_3$. Use the labeled graduated cylinders next to each of the reagent bottles for measuring out these solutions. Use these reagent solutions in your experiment.

Label five regular test tubes 1 to 5, either with labels or by noting their positions in your test tube rack. Into the first four tubes pipet 5.00 mL of 0.0120 M Pb(NO$_3$)$_2$ in KNO$_3$. Then, to test tube 1, add 2.00 mL 0.0300 M KI in KNO$_3$. Add 3, 4, and 5 mL of that same solution to test tubes 2, 3, and 4, respectively. Add enough 0.20 M KNO$_3$ to the first three test tubes to make the total volume 10.00 mL in each tube. The composition of the final mixture in each tube is summarized in Table 24.1.

Table 24.1   Volumes of Reagents Used in Precipitating PbI$_2$(mL)

Test Tube	0.0120 M Pb(NO$_3$)$_2$	0.0300 M KI	0.20 M KNO$_3$
1	5.00	2.00	3.00
2	5.00	3.00	2.00
3	5.00	4.00	1.00
4	5.00	5.00	0.00
5	10.00	10.00	0.00

Stopper each test tube and shake thoroughly at intervals of several minutes while you are proceeding with the next part of the experiment.

In the fifth test tube mix about 10 mL of 0.0120 M Pb(NO$_3$)$_2$ in KNO$_3$ with 10 mL of 0.0300 M KI in KNO$_3$. Shake the mixture vigorously for a minute or so. Let the solid settle for a few minutes and then decant 3/4 of the supernatant solution into a beaker. Transfer the solid PbI$_2$ and the rest of the solution into a small test tube and centrifuge. Pour the liquid into the same beaker, retaining the solid precipitate. Add 3 mL 0.20 M KNO$_3$ and shake to wash the solid free of excess Pb^{2+} or I$^-$ ions. Centrifuge again, and discard the liquid into the beaker. By this procedure you should now have in the test tube a small sample of essentially pure PbI$_2$ in a little KNO$_3$ solution. Add 0.20 M KNO$_3$ to the solid until the tube is about 3/4 full. Shake well at several 1-minute intervals to saturate the solution with PbI$_2$.

In this experiment it is essential that the volumes of reagents used to make up the mixtures in test tubes 1 to 4 be measured accurately. It is also essential that all five mixtures be shaken thoroughly so that equilibrium can be established. Insufficient shaking of the first four test tubes will result in not enough PbI$_2$ precipating to reach true equilibrium; if the small test tube is not shaken sufficiently, not enough PbI$_2$ will dissolve to attain equilibrium.

When each of the mixtures has been shaken for at least 15 minutes, let the tubes stand for 3 to 4 minutes to let the solid settle. Pour the supernatant liquid in test tube 1 into a small dry test tube until it is ¾ full and centrifuge for about 3 minutes to settle the solid PbI$_2$. Pour the liquid into another small dry test tube; if there are any solid particles or yellow color remaining in the liquid, centrifuge again. When you have a clear liquid, dip a small piece of clean, dry paper towel into the liquid to remove floating PbI$_2$ particles from the surface. Pipet 3.00 mL of 0.02 M KNO$_2$, potassium NITRITE (*not* KNO$_3$, potassium nitrate), into a clean, dry spectrophotometer tube and add two drops 6 M HCl. Then, using a medicine dropper, add enough of the clear centrifuged solution (about 3 mL) to fill the spectrophotometer tube just to the level indicated on the tube. Shake gently to mix the reagents and then measure the Absorbance of the solution as directed by your instructor. The calibration curve or equation that is provided will allow you to determine directly the concentration of I$^-$ ion that was in equilibrium with PbI_2. (See Appendix IV for a discussion of Absorbance.) Use the same procedure to analyze the solutions in test tubes 2 through 5, completing each analysis before you proceed to the next.

DISPOSAL OF REACTION PRODUCTS   Lead ion is toxic, as is PbI_2. When you are finished with the experiment, pour all of the solutions and solids in the test tubes and beakers into the waste crock.

Data and Calculations: Determination of the Solubility Product of PbI$_2$

From the experimental data we obtain [I$^-$] directly. To obtain K_{sp} for PbI$_2$ we must calculate [Pb^{2+}] in each equilibrium system. This is most easily done by constructing an equilibrium table. We first find the initial amounts of I$^-$ and Pb^{2+} ions in each system from the way the mixtures were made up. Knowing [I$^-$] and the formula of lead iodide allows us to calculate [Pb^{2+}]. K_{sp} then follows directly. The calculations are similar to those in Experiment 21.

$$PbI_2(s) \rightleftharpoons Pb^{2+}(aq) + 2I^-(aq) \quad K_{sp} = [Pb^{2+}] [I^-]^2$$

Data

Test tube no.	1	2	3	4	5
					Saturated soln. of PbI$_2$
mL 0.0120 M Pb(NO$_3$)$_2$	_____	_____	_____	_____	
mL 0.0300 M KI	_____	_____	_____	_____	
mL 0.20 M KNO$_3$	_____	_____	_____	_____	
Total volume in mL	_____	_____	_____	_____	
Absorbance of solution	_____	_____	_____	_____	_____
[I$^-$] in moles/liter at equilibrium	_____	_____	_____	_____	_____

Calculations

Initial no. moles Pb^{2+}	___ $\times 10^{-5}$	___ $\times 10^{-5}$	___ $\times 10^{-5}$	___ $\times 10^{-5}$
Initial no. moles I$^-$	___ $\times 10^{-5}$	___ $\times 10^{-5}$	___ $\times 10^{-5}$	___ $\times 10^{-5}$
No. moles I$^-$ at equilibrium	___ $\times 10^{-5}$	___ $\times 10^{-5}$	___ $\times 10^{-5}$	___ $\times 10^{-5}$
No. moles I$^-$ precipitated	___ $\times 10^{-5}$	___ $\times 10^{-5}$	___ $\times 10^{-5}$	___ $\times 10^{-5}$
No. moles Pb^{2+} precipitated	___ $\times 10^{-5}$	___ $\times 10^{-5}$	___ $\times 10^{-5}$	___ $\times 10^{-5}$

(continued on following page)

(continued)

No. moles Pb^{2+} at
equilibrium   ___ $\times 10^{-5}$ ___ $\times 10^{-5}$ ___ $\times 10^{-5}$ ___ $\times 10^{-5}$

$[Pb^{2+}]$ at equilibrium   _____   _____   _____   _____   _____

$K_{sp} PbI_2$ _____   _____   _____   _____   _____

Advance Study Assignment: Determination of the Solubility Product of PbI$_2$

1. State in words the meaning of the solubility product equation for PbI$_2$:

$$K_{sp} = [Pb^{2+}][I^-]^2$$

2. When 5.00 mL of 0.0120 M Pb(NO$_3$)$_2$ are mixed with 5.00 mL of 0.0300 M KI, a yellow precipitate of PbI$_2$(s) forms.

   a. How many moles of Pb^{2+} are initially present? (no. moles A = M_A × volume in liters).

   _____ moles

   b. How many moles of I$^-$ are originally present?

   _____ moles

   c. In a colorimeter the equilibrium solution is analyzed for I$^-$, and its concentration is found to be 8.0 × 10^{-3} mole/liter. How many moles of I$^-$ are present in the solution (10 mL)?

   _____ moles

   d. How many moles of I$^-$ precipitated? (You know how many you started with and how many remain in solution, so?)

   _____ moles

   e. How many moles of Pb^{2+} precipitated? (Note that PbI$_2$ is the precipitate, so the amounts of Pb^{2+} and I$^-$ that precipitated must be related.)

   _____ moles

   f. How many moles of Pb^{2+} are left in solution?

   _____ moles

   g. What is the concentration of Pb^{2+} in the equilibrium solution? (Still 10 mL)

   _____ moles/liter

   h. Find a value for K_{sp} of PbI$_2$ from these data.

   _____

(continued on following page)

3. In another experiment a small sample of pure PbI_2 is shaken with water to produce a saturated solution, as in Test tube 5.

   a. What must be the relationship between $[Pb^{2+}]$ and $[I^-]$ in that solution? (See Reaction 1.)

   b. The concentration of I^- in the saturated solution is found to be 5.0×10^{-3} M. What is the concentration of Pb^{2+} ion?

   _____ M

   c. Using the results of Part b, calculate a value for K_{sp} for PbI_2. (Use Eq. 1.)

   _____

RELATIVE STABILITIES OF COMPLEX IONS AND PRECIPITATES PREPARED FROM SOLUTIONS OF COPPER(II)

In aqueous solution, typical cations, particularly those produced from atoms of the transition metals, do not exist as free ions but rather consist of the metal ion in combination with some water molecules. Such cations are called complex ions. The water molecules, usually two, four, or six in number, are bound chemically to the metallic cation, but often rather loosely, with the electrons in the chemical bonds being furnished by one of the unshared electron pairs from the oxygen atoms in the H_2O molecules. Copper ion in aqueous solution may exist as $Cu(H_2O)_4^{2+}$, with the water molecules arranged in a square around the metal ion at the center.

If a hydrated cation such as $Cu(H_2O)_4^{2+}$ is mixed with other species that can, like water, form coordinate covalent bonds with Cu^{2+}, those species, called ligands, may displace one or more H_2O molecules and form other complex ions containing the new ligands. For instance, NH_3, a reasonably good coordinating species, may replace H_2O from the hydrated copper ion, $Cu(H_2O)_4^{2+}$, to form $Cu(H_2O)_3NH_3^{2+}$, $Cu(H_2O)_2(NH_3)_2^{2+}$, $Cu(H_2O)(NH_3)_3^{2+}$, or $Cu(NH_3)_4^{2+}$. At moderate concentrations of NH_3, essentially all the H_2O molecules around the copper ion are replaced by NH_3 molecules, forming the copper ammonia complex ion.

Coordinating ligands differ in their tendencies to form bonds with metallic cations, so that in a solution containing a given cation and several possible ligands, an equilibrium will develop in which most of the cations are coordinated with those ligands with which they form the most stable bonds. There are many kinds of ligands, but they all share the common property that they possess an unshared pair of electrons that they can donate to form a coordinate covalent bond with a metal ion. In addition to H_2O and NH_3, other uncharged coordinating species include CO and ethylenediamine; some common anions that can form complexes include OH^-, Cl^-, CN^-, SCN^-, and $S_2O_3^{2-}$.

As you know, when solutions containing metallic cations are mixed with other solutions containing ions, precipitates are sometimes formed. When a solution of 0.1 M copper nitrate is mixed with a little 1 M NH_3 solution, a precipitate forms and then dissolves in excess ammonia. The formation of the precipitate helps us to understand what is occurring as NH_3 is added. The precipitate is hydrous copper hydroxide, formed by reaction of the hydrated copper ion with the small amount of hydroxide ion present in the NH_3 solution. The fact that this reaction occurs means that even at very low OH^- ion concentration $Cu(OH)_2(H_2O)_2(s)$ is a more stable species than $Cu(H_2O)_4^{2+}$ ion.

Addition of more NH_3 causes the solid to redissolve. The copper species then in solution cannot be the hydrated copper ion. (Why?) It must be some other complex ion and is, indeed, the $Cu(NH_3)_4^{2+}$ ion. The implication of this reaction is that the $Cu(NH_3)_4^{2+}$ ion is also more stable in NH_3 solution than is the hydrated copper ion. To deduce in addition that the copper ammonia complex ion is also more stable in general than $Cu(OH)_2(H_2O)_2(s)$ is not warranted, since under the conditions in the solution $[NH_3]$ is much larger than $[OH^-]$, and given a higher concentration of hydroxide ion, the solid hydrous copper hydroxide might possibly precipitate even in the presence of substantial concentrations of NH_3.

To resolve this question, you might proceed to add a little 1 M NaOH solution to the solution containing the $Cu(NH_3)_4^{2+}$ ion. If you do this you will find that $Cu(OH)_2(H_2O)_2(s)$

does indeed precipitate. We can conclude from these observations that $Cu(OH)_2(H_2O)_2(s)$ is more stable than $Cu(NH_3)_4{}^{2+}$ in solutions in which the ligand concentrations (OH^- and NH_3) are roughly equal.

The copper species that will be present in a system depends, as we have just seen, on the conditions in the system. We cannot say in general that one species will be more stable than another; the stability of a given species depends in large measure on the kinds and concentrations of other species that are also present with it.

Another way of looking at the matter of stability is through equilibrium theory. Each of the copper species we have mentioned can be formed in a reaction between the hydrated copper ion and a complexing or precipitating ligand; each reaction will have an associated equilibrium constant, which we might call a formation constant for that species. The pertinent formation reactions and their constants for the copper species we have been considering are listed here:

$$Cu(H_2O)_4{}^{2+}(aq) + 4\,NH_3(aq) \rightleftharpoons Cu(NH_3)_4{}^{2+}(aq) + 4\,H_2O \quad K_1 = 5 \times 10^{12} \quad (1)$$

$$Cu(H_2O)_4{}^{2+}(aq) + 2\,OH^-(aq) \rightleftharpoons Cu(OH)_2(H_2O)_2(s) + 2\,H_2O \quad K_2 = 2 \times 10^{19} \quad (2)$$

The formation constants for these reactions do not involve $[H_2O]$ terms, which are essentially constant in aqueous systems and are included in the magnitude of K in each case. The large size of each formation constant indicates that the tendency for the hydrated copper ion to react with the ligands listed is very high.

In terms of these data, let us compare the stability of the $Cu(NH_3)_4{}^{2+}$ complex ion with that of solid $Cu(OH)_2(H_2O)_2$. This is most readily done by considering the reaction:

$$Cu(NH_3)_4{}^{2+}(aq) + 2\,OH^-(aq) + 2\,H_2O \rightleftharpoons Cu(OH)_2(H_2O)_2(s) + 4\,NH_3(aq) \quad (3)$$

We can find the value of the equilibrium constant for this reaction by noting that it is the sum of Reaction 2 and the reverse of Reaction 1. By the Law of Multiple Equilibrium, K for Reaction 3 is given by the equation

$$K = \frac{K_2}{K_1} = \frac{2 \times 10^{19}}{5 \times 10^{12}} = 4 \times 10^6 = \frac{[NH_3]^4}{[Cu(NH_3)_4{}^{2+}][OH^-]^2} \quad (4)$$

From the expression in Equation 4 we can calculate that in a solution in which the NH_3 and OH^- ion concentrations are both about 1 M,

$$[Cu(NH_3)_4{}^{2+}] = \frac{1}{4 \times 10^6} = 2.5 \times 10^{-7}\,M \quad (5)$$

Since the concentration of the copper ammonia complex ion is very, very low, any copper(II) in the system will exist as the solid hydroxide. In other words, the solid hydroxide is more stable under such conditions than the ammonia complex ion. But that is exactly what we observed when we treated the hydrated copper ion with ammonia and then with an equivalent amount of hydroxide ion.

Starting now from the experimental behavior of the copper ion, we can conclude that since the solid hydroxide is the species that exists when copper ion is exposed to equal concentrations of ammonia and hydroxide ion, the hydroxide is more stable under those conditions, *and* the equilibrium constant for the formation of the hydroxide is larger than the constant for the formation of the ammonia complex. By determining, then, which species is present when a cation is in the presence of equal ligand concentrations, we can speak meaningfully of stability under such conditions and can rank the formation constants for the possible complex ions, and indeed for precipitates, in order of their increasing magnitudes.

In this experiment you will carry out formation reactions for a group of complex ions and precipitates involving the Cu^{2+} ion. You can make these species by mixing a solution of $Cu(NO_3)_2$ with solutions containing NH_3 or anions, which may form either precipitates or complex ions by reaction with $Cu(H_2O)_4^{2+}$, the cation present in aqueous solutions of copper(II) nitrate. By examining whether the precipitates or complex ions formed by the reaction of hydrated copper(II) ion with a given species can, on addition of a second ligand, be dissolved or transformed to another species, you will be able to rank the relative stabilities of the precipitates and complex ions made from Cu^{2+} with respect to one another, and thus rank the equilibrium formation constants for each species in order of increasing magnitude. The species to be reacted with Cu^{2+} ion in aqueous solution are NH_3, Cl^-, OH^-, $C_2O_4^{2-}$, S^{2-}, NO_2^-, and PO_4^{3-}. In each case the test for relative stability will be made in the presence of essentially equal concentrations of the two ligands. When you have completed your ranking of the known species you will test an unknown species and incorporate it into your list.

Experimental Procedure

Obtain from the stockroom an unknown and eight small test tubes.

Add about 1 mL 0.1 M $Cu(NO_3)_2$ solution to each of the test tubes (1/2-inch depth).

To one of the test tubes add about 1 mL 1 M NH_3, drop by drop. Note whether a precipitate forms initially, and if it dissolves in excess NH_3. Shake the tube sideways to mix the reagents. In the NH_3-NH_3 space in the table (Data page), write a P in the upper left-hand corner if a precipitate is formed initially. In the rest of the space, write the formula and the color of the species that was present with an excess of NH_3. If a solution is present, the copper ion will be in a complex. Cu(II) will always have a coordination number of 4, so the formula with NH_3 would be $Cu(NH_3)_4^{2+}$. If a precipitate is present, it will be neutral, and in the case of NH_3 it would be a hydroxide with the formula $Cu(OH)_2$. (There should in principle be two H_2O molecules in the formula, but they are usually omitted.) Add 1 mL 1 M NH_3 to the rest of the test tubes.

Now you will test the stability of the species present in excess NH_3 relative to those which might be present with other precipitating or coordinating species. Add, drop by drop, 1 mL of a 1 M solution of each of the anions in the horizontal row of the table to the test tubes you have just prepared, one solution to a test tube. Note any changes that occur in the appropriate spaces in the table. A change in color or the formation of a precipitate implies that a reaction has occurred between the added ligand or precipitating anion and the species originally present. As before, put a P in the upper left-hand corner if a precipitate initially forms on addition of the anion solution. In the rest of the space, write the formula and color of the species present when the anion is in excess. That is the species that is stable in the presence of equal concentrations of NH_3 and the added anion. Again, in complex ions, Cu(II) will normally be attached to four ligands; copper(II) precipitates will be neutral. If a new species forms on addition of the second reagent, its formula should be given. If no change occurs, the original species is more stable, so put its formula in that space.

Repeat the above series of experiments, using 1 M Cl^- as the species originally added to the $Cu(NO_3)_2$ solution. In each case record the color of any precipitates or solutions formed on addition of the reagents in the horizontal row, and the formula of the species stable when an excess of both Cl^- ion and the added species is present in the solution. Since these reactions are reversible, it is not necessary to retest Cl^- with NH_3, since the same results would be obtained as when the NH_3 solution was tested with Cl^- solution.

Repeat the series of experiments for each of the anions in the vertical row in the table, omitting those tests where decisions as to relative stabilities are already clear. Where both ligands produce precipitates it will be helpful to check the effect of the addition of the other ligand to those precipitates. When complete, your table should have at least 28 entries.

Examine your table and decide on the relative stabilities of all species you observed to be present in all the spaces of the table. There should be seven such species; rank them as best you can in order of increasing stability. There is only one correct ranking, and you should be prepared to defend your choices. Although we did not in general prepare the species by direct reaction of $Cu(H_2O)_4^{2+}$ with the added ligand or precipitating anion, the equilibrium formation constants for those species for the direct reactions will have magnitudes that increase in the same order as the relative stabilities of the species you have established.

When you are satisfied that your ranking order is correct, carry out the necessary tests on your unknown to determine its proper position in the list. Your unknown may be one of the Cu(II) species you have already observed, or it may be a different species, present in excess of its ligand or precipitating anion. If your unknown contains a precipitate, shake it well before using a portion of it to make a test.

Dispose of all reaction products in the waste crock provided or as otherwise directed by your instructor.

Alternate Procedure Using Microscale (Optional)

From the stockroom obtain an unknown and two plastic well plates (4 × 6).

Align the plates into a 6 × 8 well configuration, with six wells across the top. To each of the six wells in the top row of the plate add six drops of 0.1 M $Cu(NO_3)_2$. Use a Beral pipette for this and all other additions of reagents. To the six wells add six drops of 1 M solutions containing the species in the horizontal row at the top of the Table of Observations; one species to a well, starting with 1 M NH_3 in the first well and ending with 1 M $NaNO_2$ in the sixth (we will omit S^{2-} and come back to it later). Mix the reagents by moving the well plate back and forth on the top of the lab bench. You will notice a difference in the appearance of the mixtures in each of the six wells. Each contains the species that is stable when Cu(II) is mixed with one of the reagents. Report the color and the formula of those six species along the **diagonal** blocks in the Table, using the directions in the first two paragraphs in the ordinary procedure. The contents of these wells will serve as a reference during the rest of the experiment.

To establish the relative stabilities of the six species, we need to mix them with each of the other species in some systematic way and see what happens. We will first test them against NH_3. To do this, in the second row of wells make the same mixtures as you did in the first row. Then, add six drops of 1 M NH_3 to the second through sixth wells in that row (we do not need to test NH_3 with NH_3, since we really have already done that). Mix the reagents, and note any changes that occurred on addition of NH_3. In some of the wells, changes will occur, indicating a reaction to form a more stable species has gone on. Use the wells in the top row for comparison. Record your observations in the blocks in the first row of the Table, noting the color and formula of the stable Cu(II) species in each case.

Now make stability tests for Cl^-, using wells in the third row of the plate. This time start with the third well, since we already know how Cl^- and NH_3 stack up from the last set of tests. (It doesn't make any difference if we change the order in which the reagents are added.) The four wells on the right end of the third row should be prepared in the same way as the corresponding wells in the first row. To each of those wells add six drops of 1 M NaCl, mix, note any changes, and report your observations as before in the blocks in the second row of the Table.

Continue along these lines, making stability tests with 1 M NaOH in the bottom row of the top plate. Report your observations as before. Then, in the top three rows of the lower well plate, make the same sort of tests with 1 M $K_2C_2O_4$, 1 M Na_2HPO_4, and 1 M KNO_2, respectively. When you are done, you should have entries in all of the blocks in the upper right half of the Table except for those involving S^{2-}.

To establish the stability of the Cu(II) species with respect to the sulfide ion, S^{2-}, add a drop or two of 0.1 M $(NH_4)_2S$ to each of the wells in the top row of the top plate. Although the sulfide solution is dilute, its effect on the contents of the wells, and on the nose, should be clear. Fill in the last column of the Table.

Given your entries in the Table, you should be able to decide on the relative stabilities of the seven Cu(II) species studied in this experiment. Repeat any tests that appear ambiguous, changing the order in which reagents are added if necessary. List the seven species, in order of increasing stability. Then carry out the necessary tests on your unknown to establish its position in the list. The unknown may be one of the Cu(II) species you have studied, or it may be different. If the unknown contains a precipitate, stir it up before making a test.

Pour the contents of all of the wells into the waste crock unless directed otherwise by your instructor.

Name _____ *Section* _____

Data and Observations: Relative Stabilities of Complex Ions and Precipitates Containing Cu(II)

Table of Observations

	NH₃	Cl⁻	OH⁻	C₂O₄²⁻	PO₄³⁻	NO₂⁻	S²⁻	Unknown
S^{2-}								
NO_2^-								
PO_4^{3-}								
$C_2O_4^{2-}$								
OH^-								
Cl^-								
NH_3								

(continued on following page)

(continued)

Determination of Relative Stabilities

In each row of the table you can compare the stabilities of species involving the reagent in the horizontal row with those of the species containing the reagent initially added. In the first row of the table, the copper(II)-NH_3 species can be seen to be more stable than some of the species obtained by addition of the other reagents, and less stable than others. Examining each row, make a list of all the complex ions and precipitates you have in the table in order of increasing stability and formation constant.

Reasons

Lowest _____   _____

_____   _____

_____   _____

_____   _____

_____   _____

Highest _____   _____

Stability of Unknown

Indicate the position your unknown would occupy in the above list.

Reasons:

Unknown no. _____

Advance Study Assignment: Stabilities of Complex Ions and Precipitates
of Cu(II)

1. In testing the relative stabilities of Cu(II) species a student adds 2 mL 1 M NH_3 to 2 mL 0.1 M $Cu(NO_3)_2$. He observes that a blue precipitate initially forms, but that in excess NH_3 the precipitate dissolves and the solution turns blue. Addition of 2 mL 1 M NaOH to the dark-blue solution results in the formation of a blue precipitate.

    a. What is the formula of Cu(II) species in the dark-blue solution?

    b. What is the formula of the blue precipitate present after addition of 1 M NaOH?

    c. Which species is more stable in equal concentrations of NH_3 and OH^-, the one in Part a or the one in Part b?

    _____

2. Given the following two reactions and their equilibrium constants:

    $$Cu(H_2O)_4{}^{2+}(aq) + 4\ NH_3(aq) \rightleftharpoons Cu(NH_3)_4{}^{2+}(aq) + 4\ H_2O \qquad K_1 = 5 \times 10^{12}$$

    $$Cu(H_2O)_4{}^{2+}(aq) + CO_3{}^{2-} \rightleftharpoons CuCO_3(s) + 4\ H_2O \qquad K_2 = 7 \times 10^9$$

    a. Evaluate the equilibrium constant for the reaction (see discussion):

    $$CuCO_3(s) + 4\ NH_3(aq) \rightleftharpoons Cu(NH_3)_4{}^{2+}(aq) + CO_3{}^{2-}(aq)$$

    _____

    b. If 1 M NH_3 were added to some solid $CuCO_3$ in a test tube containing 1 M Na_2CO_3, what, if anything, would happen? Explain your reasoning.

xperiment 26

DETERMINATION OF THE HARDNESS OF WATER

One of the factors that establishes the quality of a water supply is its degree of hardness. The hardness of water is defined in terms of its content of calcium and magnesium ions. Since the analysis does not distinguish between Ca^{2+} and Mg^{2+}, and since most hardness is caused by carbonate deposits in the earth, hardness is usually reported as total parts per million calcium carbonate by weight. A water supply with a hardness of 100 parts per million would contain the equivalent of 100 grams of $CaCO_3$ in 1 million grams of water or 0.1 gram in one liter of water. In the days when soap was more commonly used for washing clothes, and when people bathed in tubs instead of using showers, water hardness was more often directly observed than it is now, since Ca^{2+} and Mg^{2+} form insoluble salts with soaps and make a scum that sticks to clothes or to the bathtub. Detergents have the distinct advantage of being effective in hard water, and this is really what allowed them to displace soaps for laundry purposes.

Water hardness can be readily determined by titration with the chelating agent EDTA (ethylenediaminetetraacetic acid). This reagent is a weak acid that can lose four protons on complete neutralization; its structural formula is

$$
\begin{array}{cc}
HOOC-CH_2 & CH_2-COOH \\
\quad \diagdown & \diagup \\
\quad N-CH_2-CH_2-N \\
\quad \diagup & \diagdown \\
HOOC-CH_2 & CH_2-COOH
\end{array}
$$

The four acid sites and the two nitrogen atoms all contain unshared electron pairs, so that a single EDTA ion can form a complex with up to six sites on a given cation. The complex is typically quite stable, and the conditions of its formation can ordinarily be controlled so that it contains EDTA and the metal ion in a 1 : 1 mole ratio. In a titration to establish the concentration of a metal ion, the EDTA which is added combines quantitatively with the cation to form the complex. The end point occurs when essentially all of the cation has reacted.

In this experiment we will standardize a solution of EDTA by titration against a standard solution made from calcium carbonate, $CaCO_3$. We will then use the EDTA solution to determine the hardness of an unknown water sample. Since both EDTA and Ca^{2+} are colorless, it is necessary to use a rather special indicator to detect the end point of the titration. The indicator we will use is called Eriochrome Black T, which forms a rather stable wine-red complex, $MgIn^-$, with the magnesium ion. A tiny amount of this complex will be present in the solution during the titration. As EDTA is added, it will complex free Ca^{2+} and Mg^{2+} ions, leaving the $MgIn^-$ complex alone until essentially all of the calcium and magnesium have been converted to chelates. At this point EDTA concentration will increase sufficiently to displace Mg^{2+} from the indicator complex; the indicator reverts to an acid form, which is sky blue, and this establishes the end point of the titration.

The titration is carried out at a pH of 10, in an $NH_3-NH_4^+$ buffer, which keeps the EDTA (H_4Y) mainly in the form, HY^{3-}, where it complexes the Group 2 ions very well but does not tend to react as readily with other cations such as Fe^{3+} that might be present as impurities in the water. Taking H_4Y and H_3In as the formulas for EDTA and Eriochrome Black T respectively, the equations for the reactions which occur during the titration are:

(main reaction) $HY^{3-}(aq) + Ca^{2+}(aq) \rightarrow CaY^{2-}(aq) + H^+(aq)$ (same for Mg^{2+})

(at end point) $HY^{3-}(aq) + MgIn^-(aq) \rightarrow MgY^{2-}(aq) + HIn^{2-}(aq)$
                             wine red   sky blue

Since the indicator requires a trace of Mg^{2+} to operate properly, we will add a little magnesium ion to each solution and titrate it as a blank.

Experimental Procedure

Obtain a 50-mL buret, a 250-mL volumetric flask, and 25- and 50-mL pipets from the stockroom.

Put about a half gram of calcium carbonate in a small 50-mL beaker and weigh the beaker and contents on the analytical balance. Using a spatula, transfer about 0.4 g of the carbonate to a 250-mL beaker and weigh again, determining the mass of the $CaCO_3$ sample by difference.

Add 25 mL of distilled water to the large beaker and then, *slowly,* about 40 drops of 6 M HCl. Cover the beaker with a watch glass and allow the reaction to proceed until all of the solid carbonate has dissolved. Rinse the walls of the beaker down with distilled water from your wash bottle and heat the solution until it just begins to boil. (Be sure not to be confused by the evolution of CO_2 which occurs with the boiling.) Add 50 mL of distilled water to the beaker and carefully transfer the solution, using a stirring rod as a pathway, to the volumetric flask. Rinse the beaker several times with small portions of distilled water and transfer each portion to the flask. All of the Ca^{2+} originally in the beaker should then be in the volumetric flask. Fill the volumetric flask to the horizontal mark with distilled water, adding the last few mL a drop at a time with your wash bottle. Stopper the flask and mix the solution thoroughly by inverting the flask at least 20 times over a period of several minutes.

Clean your buret thoroughly and draw about 200 mL of the stock EDTA solution from the carboy into a dry 250-mL Erlenmeyer flask. Rinse the buret with a few mL of the solution at least three times. Drain through the stopcock and then fill the buret with the EDTA solution.

Determine a blank by adding 25 mL distilled water and 5 mL of the pH 10 buffer to a 250-mL Erlenmeyer flask. Add a small amount of solid Eriochrome Black T indicator mixture from the stock bottle. You need only a small portion, about 25 mg, just enough to cover the end of a small spatula. The solution should turn blue; if the color is weak, add a bit more indicator. Add 15 drops 0.03 M $MgCl_2$, which should contain enough Mg^{2+} to turn the solution wine red. Read the buret to 0.02 mL and add EDTA to the solution until the last tinge of purple just disappears. The color change is rather slow, so titrate slowly near the end point. Only a few mL will be needed to titrate the blank. Read the buret again to determine the volume required for the blank. This volume must be subtracted from the total EDTA volume used in each titration. Save the solution as a reference for the end point in all your titrations.

Pipet three 25-mL portions of the Ca^{2+} solution in the volumetric flask into clean 250-mL Erlenmeyer flasks. To each flask add 5 mL of the pH 10 buffer, a small amount of indicator, and 15 drops of 0.03 M $MgCl_2$. Titrate the solution in one of the flasks until its color matches that of your reference solution; the end point is a reasonably good one, and you should be able to hit it within a few drops if you are careful. Read the buret. Refill the buret, read it, and titrate the second solution, then the third.

Your instructor will furnish you a sample of water for hardness analysis. Since the concentration of Ca^{2+} is probably lower than that in the standard calcium solution you prepared, pipet 50 mL of the water sample for each titration. As before, add some indicator, 5 mL of pH 10 buffer, and 15 drops of 0.03 M $MgCl_2$ before titrating. Carry out as many titrations as necessary to obtain two volumes of EDTA that agree within about 3%. If the volume of EDTA required in the first titration is low due to the fact that the water is not very hard, increase the volume of the water sample so that in succeeding titrations, it takes at least 20 mL of EDTA to reach the end point.

The chemical waste from this experiment may be diluted with water and poured down the sink.

Data and Calculations: Determination of the Hardness of Water

Mass of beaker
plus $CaCO_3$ _____ g

Volume Ca^{2+}
solution prepared   _____ mL

Mass of beaker
less sample   _____ g

Molarity of Ca^{2+} _____ *M*

Mass of $CaCO_3$
sample   _____ g

Moles Ca^{2+} in each
aliquot titrated   _____ moles

Number of moles $CaCO_3$ in sample
(Formula mass = 100.1)   _____ moles

Standardization of EDTA Solution

Determination of blank:

Initial buret
reading   _____ mL

Final buret
reading   _____ mL

Volume of
blank   _____ mL

Titration:	*I*	*II*	*III*
Initial buret reading	_____ mL	_____ mL	_____ mL
Final buret reading	_____ mL	_____ mL	_____ mL
Volume of EDTA	_____ mL	_____ mL	_____ mL
Volume EDTA used to titrate blank	_____ mL	_____ mL	_____ mL
Volume EDTA used to titrate Ca^{2+}	_____ mL	_____ mL	_____ mL

Average volume of EDTA required to titrate Ca^{2+} _____ mL

$$\text{Molarity of EDTA} = \frac{\text{no. moles } Ca^{2+} \text{ in aliquot} \times 1000}{\text{average volume EDTA required (mL)}} = \underline{\hspace{2cm}} \, M$$

(continued on following page)

(continued)

Determination of Water Hardness

Titration:	*I*	*II*	*III*
Volume of water used	_____ mL	_____ mL	_____ mL
Initial buret reading	_____ mL	_____ mL	_____ mL
Final buret reading	_____ mL	_____ mL	_____ mL
Volume of EDTA	_____ mL	_____ mL	_____ mL
Volume of EDTA used to titrate blank	_____ mL	_____ mL	_____ mL
Volume of EDTA required to titrate water	_____ mL	_____ mL	_____ mL

Average volume EDTA
per liter of water _____ mL

No. moles EDTA per liter water _____ = No. moles $CaCO_3$ per liter water

No. grams $CaCO_3$
per liter water _____ g

Water hardness
(1 ppm = 1 mg/liter) _____ ppm $CaCO_3$

Unknown No. _____

Advance Study Assignment: Determination of the Hardness of Water

1. A 0.5215-g sample of $CaCO_3$ is dissolved in 12 M HCl and the resulting solution is diluted to 250.0 mL in a volumetric flask.

   a. How many moles of $CaCO_3$ are used? (formula mass = 100.1)

   _____ moles

   b. What is the molarity of the Ca^{2+} in the 250 mL of solution?

   _____ M

   c. How many moles of Ca^{2+} are in a 25.0-mL aliquot of the solution in 1b?

   _____ moles

2. 25.00-mL aliquots of the solution from Problem 1 are titrated with EDTA to the Eriochrome Black T end point. A blank containing a small measured amount of Mg^{2+} requires 2.60 mL of the EDTA to reach the end point. An aliquot to which the same amount of Mg^{2+} is added requires 28.55 mL of the EDTA to reach the end point.

   a. How many milliliters of EDTA are needed to titrate the Ca^{2+} ion in the aliquot?

   _____ mL

   b. How many moles of EDTA are there in the volume obtained in Part a?

   _____ moles

   c. What is the molarity of the EDTA solution?

   _____ M

3. A 100-mL sample of hard water is titrated with the EDTA solution in Problem 2. The same amount of Mg^{2+} is added as previously, and the volume of EDTA required is 22.44 mL.

   a. What volume of EDTA is used in titrating the Ca^{2+} in the hard water?

   _____ mL

   b. How many moles of EDTA are there in that volume?

   _____ moles

   c. How many moles of Ca^{2+} are there in the 100 mL of water?

   _____ moles

(continued on following page)

239

(continued)

d. If the Ca^{2+} comes from $CaCO_3$, how many moles of $CaCO_3$ are there in one liter of the water? How many grams $CaCO_3$ per liter?

_____ moles

_____ g

e. If 1 ppm $CaCO_3$ = 1 mg per liter, what is the water hardness in ppm $CaCO_3$?

_____ ppm $CaCO_3$

xperiment 27

SYNTHESIS AND ANALYSIS OF A COORDINATION COMPOUND

Some of the most interesting research in inorganic chemistry has involved the preparation and properties of coordination compounds. These compounds, sometimes called *complexes,* are typically salts that contain complex ions. A complex ion contains a central metal atom to which are bonded small polar molecules or anions, called *ligands.* The bonding in the complex is through coordinate covalent bonds, bonds in which the electrons are all furnished by the ligands.

Some coordination compounds can be prepared stoichiometrically pure. That is, the atoms in the compound are present in the exact ratio given by the formula. Many supposedly pure compounds do not meet this criterion, since they contain variable amounts of water, either in hydrates or adsorbed. In this experiment we will be making a compound that is indeed stoichiometrically pure. It is not a hydrate nor does it tend to adsorb water.

The compound we will be preparing contains cobalt ion, ammonia, and chloride ions. Its formula is of the form $Co_x(NH_3)_yCl_z$. Once you have made the compound, you may, depending on the time available, analyze it for its content of one or more of these species and so determine x, y, and/or z.

In many reactions involving complex ion formation the rate of reaction is very fast, so that the thermodynamically stable form of the complex is the one produced. By changing ligand concentrations, one can quickly convert from one complex to another. In Experiment 25, in which many Cu(II) complexes are prepared, all of the complex ions undergo rapid reaction, exchanging ligands very readily. Such complexes are called *labile.* With Cu(II) ion and NH_3 in solution, one can, depending on the NH_3 concentration, have one, two, three, or four NH_3 molecules in the complex, in an equilibrium mixture.

Most, but by no means all, complexes are labile. Some, including the one you will be making, exchange ligands only very slowly. For such species, the preparation reaction takes time, but once formed, the complex ion in solution may resist change rather dramatically. Such complexes are called *inert.*

In the complex ion you will be making, NH_3 and Cl^- may be ligands. Since the overall charge of the compound must be zero, the number of moles of Cl^- ion in a mole of compound will be determined by the charge on the cobalt ion. Some of the Cl^- ions may be in the complex and some may simply be anions in the crystal. When the compound is dissolved in water, the free anions will go into solution, but the ones in the complex will remain firmly attached to cobalt. They will not react with a precipitating agent such as Ag^+ ion, which will tend to form AgCl with any free Cl^- ions. All of the NH_3 molecules will be in the complex ion, and will not react with added H^+ ion, as they would if they were free. To ensure that we find all the Cl^- and NH_3 that is present in the compound, we will destroy the complex before attempting any analyses.

The complete analysis involves a gravimetric procedure for chloride ion, a colorimetric method for cobalt ion, and a volumetric procedure for ammonia. In some of the earlier experiments we have used these techniques (Exps. 4, 6, 22, and 26), so you may be somewhat familiar with them. From the analyses we can determine the number of moles of each species present in a mole of compound. Because the mole ratio must equal the atom or molecule ratio,

241

we can determine the formula of the compound. Although we describe the procedures for analysis of all three species in the compound, the formula can be determined from only two analyses, since the amount of the third species can be found by difference.

The synthesis of the coordination compound will take one lab period. The analysis for both chloride and cobalt content can be done in one period. To find the amount of ammonia will take most of one period.

Experimental Procedure

A. Preparation of $Co_x(NH_3)_yCl_z$

Using a piece of weighing paper and a top-loading balance, weigh out 10 ± 0.2 g of ammonium chloride, NH_4Cl. Pour the NH_4Cl into a 250-mL beaker and add 40 mL distilled water. To this add 8 ± 0.2 g of cobalt(II) chloride hexahydrate, $CoCl_2 \cdot 6 H_2O$ and 0.8 g activated charcoal (Norite). Heat this mixture to the boiling point, stirring to dissolve the soluble components.

Cool the beaker in water, then in an ice bath. Slowly, in a hood, add 40 mL 15 M NH_3, ammonia. (**CAUTION:** *This is a concentrated reagent, with a strong NH_3 odor.*) After stirring, add 50 mL 10% H_2O_2, hydrogen peroxide, slowly, a few mL at a time, while continuing to stir. (**CAUTION:** *Be careful not to spill this reagent on your skin. Use gloves.*)

When the bubbling has stopped, place the 250-mL beaker in a 600-mL beaker containing 100 mL of water at about 60°C. Leave the beaker in the water bath for 30 to 40 minutes, holding the temperature of the bath at about 60°C as long as the liquid has a pink color. Stir occasionally.

While the mixture is reacting, prepare a fritted glass filter crucible. Fit the crucible on a filter flask, put in the glass disk, and wet the disk with distilled water. Clean the crucible by pulling, under suction, about 15 mL 1 M HNO_3 through the disk, followed by 50 to 75 mL of distilled water. Put the crucible in an oven at 150°C for about an hour to dry. Prepare a second crucible the same way, and put it in the oven.

When the 40 minutes are up, remove the 250-mL beaker from the water bath and cool it, first in a water bath and then in an ice bath, until the liquid in the beaker has a temperature below 5°C. Hold the mixture at this temperature for at least 5 minutes, stirring occasionally to promote crystallization of the crude product. Set up a Buchner funnel and, with suction, filter the mixture through the filter paper in the funnel. Discard the filtrate in the waste crock.

Scrape the solid from the filter paper into the 250-mL beaker. Add 100 mL distilled water and, in a hood, 4 mL 12 M HCl. (**CAUTION:** *This reagent has a choking odor.*) Heat the mixture to boiling, with stirring, to dissolve the crystals. Rinse out the suction flask with distilled water, and reassemble the Buchner funnel. Holding the beaker with a pair of tongs or a folded paper towel, filter the hot mixture through the paper in the funnel, under suction. In this operation, the charcoal is removed and should be on the filter paper. The filtrate should be golden yellow and contains the product we seek.

Transfer the filtrate to the (rinsed-out) 250-mL beaker and, in the hood, add 15 mL 12 M HCl. Place the beaker in an ice bath and stir for 5 minutes or so to promote formation of the golden crystals of the product, $Co_x(NH_3)_yCl_z$. Check to see that the temperature is below 5°C.

Pour about 50 mL of ice-cold distilled water on to a piece of filter paper in the Buchner funnel. After 2 or 3 minutes, turn on the suction and pull the water into the suction flask. Empty the flask. Then filter the cold mixture containing the product through the funnel, using suction. Turn the suction off, and pour about 20 mL 95% ethanol on to the crystals. Wait about 10 seconds, and then turn on the suction to pull through the ethanol, which should carry with it most of the water and HCl remaining on the crystals. Draw air through the crystals for several minutes. Weigh a 100-mL beaker to 0.1 g. Transfer the crystals from the filter paper to the beaker. Put the beaker in your locker. Take the fritted glass crucibles out of the oven, using tongs or a folded paper towel, and put the crucibles on a paper towel in your locker. Discard the liquid in the suction flask in the waste crock or as directed by your instructor.

B. Gravimetric Determination of Chloride Content

Weigh your prepared compound in the 100-mL beaker to 0.0001 g on an analytical balance. (If a week has gone by, the compound should be well dried. If you are attempting to proceed on the same day as you prepared the compound, you must dry it thoroughly. This could be done by putting the sample under vacuum, while keeping the sample warm. If the sample does not lose 0.0001 g in 2 minutes while on the analytical balance, you may presume that it is dry.)

Transfer 0.3 ± 0.05 g of the compound from the beaker to a labeled 250-mL beaker, and reweigh the 100-mL beaker to 0.0001 g. Transfer a second sample of about the same size to another labeled 250-mL beaker, and again weigh the 100-mL beaker and the remaining compound accurately. Record all masses on the Data page. Cover the 100-mL beaker with a watch glass and put it in your locker.

Add 25 mL of distilled water to the sample in one of the 250-mL beakers. Add 5 mL 6 M NaOH and stir the mixture until all of the solid has dissolved. Heat the beaker and its contents to boiling, and simmer for 3 minutes, stirring occasionally. This step will destroy the complex; the mixture will turn black because of formation of Co_3O_4.

Once the black Co_3O_4 is formed, turn off the heat and let the beaker cool for a few minutes. Then add 6 M HNO_3 until the mixture is acidic to litmus (it will take about 5 mL). Add 1 mL more of the HNO_3. Add a small scoop of solid Na_2SO_3, sodium sulfite (approximately 0.2 g) and stir; this will reduce the Co_3O_4 and produce the Co^{2+} ion. The solution should clear and turn pink. Wash down the walls of the beaker with water to bring any unreacted dark material into contact with the solution. Reheat to boiling and simmer gently for a minute or so. Then add 75 mL of distilled water.

Slowly, with stirring, add 50 mL 0.1 M $AgNO_3$, silver nitrate. This will precipitate all of the chloride ion in the solution as silver chloride, AgCl. Heat the mixture, with occasional stirring, to the boiling point. This will help coagulate the AgCl crystals and so facilitate filtration. Remove the beaker from the wire gauze and cover with a watch glass. Let it cool for 10 minutes or so.

While the beaker is cooling, take one of the fritted glass crucibles you prepared in the previous session from your locker and weigh it to 0.0001 g on the analytical balance. Then set up the filter crucible apparatus.

Carefully, with suction on, transfer *all* of the AgCl precipitate to the filter crucible. Use your wash bottle and rubber policeman to complete the transfer. Wash the AgCl, with suction on, with 10 mL H_2O, followed by 5 mL 6 M HNO_3, followed by 20 mL H_2O. Keep the suction on for a minute after the last liquid has been pulled through the filter.

Put the crucible and its contents in an oven at 150°C for at least an hour. Then let it cool to room temperature in a covered beaker. When it is cool, weigh it to 0.0001 g on the analytical balance. Put the solid AgCl in the AgCl crock.

Repeat the entire procedure, using the second sample of compound.

C. Colorimetric Determination of Cobalt Content

Transfer 0.5 ± 0.05 g of your compound into a clean, dry 50-mL beaker, using the same procedure as in the chloride determination. Weighings should be made to 0.0001 g. Make another transfer of about 0.5 g into another 50-mL beaker, again weighing the 100-mL beaker and its contents accurately.

Cover each beaker with a watch glass and heat one of them gently with the Bunsen burner until the solid melts, foams, and turns blue. In this process the complex is destroyed and cobalt ion freed from the ligands. Allow the beaker and its contents to cool. Repeat the procedure with the sample in the other beaker.

To the solid in the first beaker add 5 mL 6 M H_2SO_4 and 5 mL distilled water. Heat the beaker until the liquid begins to boil and the solid is all dissolved. Wash any deposits on the watch glass into the beaker with a few milliliters of water from your wash bottle. Repeat these steps with the sample in the other beaker.

Transfer the solution from the first beaker to a clean 25-mL volumetric flask. Rinse the beaker with small portions of distilled water, adding the washings to the volumetric flask, so that all of the solution goes into the flask. Fill the flask to the mark with distilled water from your wash bottle. Stopper the flask and invert it at least twenty times to ensure that the solution inside becomes thoroughly mixed.

Pour some of the solution in the flask into a spectrophotometric test tube (3/4 full). Measure and record the absorbance of the solution at 510 nm. From the absorbance and the calibration curve furnished by your instructor, determine the molarity of cobalt ion in the solution.

Pour the rest of the solution in the volumetric flask in the waste crock. Rinse the flask, and use it with the solution you prepared in the other beaker. Treat that solution as you did the first one, recording its absorbance and molarity of cobalt ion on the Data page.

D. Volumetric Determination of Ammonia Content

In this part of the experiment we will first decompose the complex, releasing NH_3 to the solution, as we did in the chloride analysis. We then distill off the NH_3 into another container, and titrate it against an acid. The procedure is called a Kjeldahl analysis, and is used for determining nitrogen content of many organic and inorganic substances.

Assemble the distillation apparatus shown in Figure 27.1. The details of the apparatus will vary from school to school but you will need a 250-mL distilling flask, a distilling head, a condenser, and a receiver adapter, in addition to a 125-mL Erlenmeyer flask, which will serve as a receiving flask. The 250-mL flask should be on a wire gauze on an iron ring. The flask and condenser should be held with clamps, which are to be adjusted so that all joints are tight. On the end of the receiver adapter attach a 4-inch length of latex tubing, which should reach to the bottom of the receiving flask.

Put 50 mL saturated boric acid solution in the 125-mL receiving flask; add five drops bromcresol green indicator, and adjust the level of the flask, if necessary, so that the latex tubing is well under the liquid surface.

Weigh out 1 ± 0.1 g of your compound into a beaker, as you did earlier, making the weighing to 0.0001 g. Disconnect the distilling flask from the head. Transfer the sample to the distilling flask, using a funnel. Rinse the beaker several times with small portions of distilled water, and add the washings to the flask. All the sample must end up in the flask. Add distilled water as necessary to give a final volume of about 50 mL. Swirl the flask to dissolve the sample. Add a few boiling chips and several pieces of granulated zinc. Start water flowing through the condenser, slowly. If you are working with standard taper glassware, lightly grease the lower joint of the distilling head.

Pour 40 mL 6 M NaOH and 50 mL distilled water into the flask and reconnect it promptly to the distilling head. Make sure that all joints are tight, and that the top of the distilling head is stoppered. Turn on the Bunsen burner and bring the liquid in the flask to a boil. The complex will be destroyed, the liquid will turn dark as Co_3O_4 forms, and NH_3 will be driven from the solution into the receiving flask.

Adjust the burner as necessary to maintain smooth boiling and minimize bumping. The color of the indicator will change as the NH_3 is absorbed. There may be a tendency for liquid to rise in the latex tubing during the distillation; this is not serious, but if the level goes up

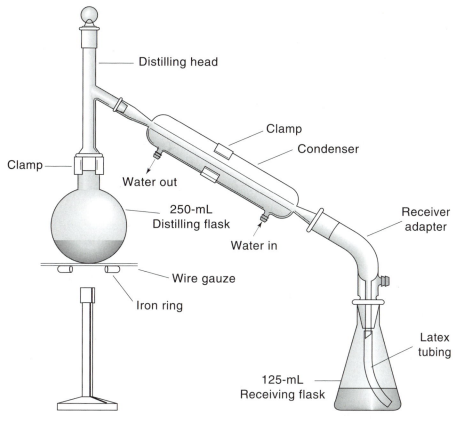

Distilling head

Clamp

Condenser

Clamp

Water out

250-mL
Distilling flask

Water in

Receiver
adapter

Wire gauze

Iron ring

Latex
tubing

125-mL
Receiving flask

FIGURE 27.1

more than a few inches, you can let in a little air by momentarily loosening one of the joints. When the volume of liquid in the receiving flask reaches 100 mL, 50 mL of the solution will have been distilled and essentially all of the NH_3 driven into the boric acid.

With the burner still heating the flask, disconnect the receiver adapter from the condenser. (If the heat is turned off first, liquid will tend to back up into the condenser.) Rinse the adapter with the tubing still in the flask, washing any liquid on the inner or outer surface into the distillate. Then turn the burner off. Pour the distillate into a 250-mL volumetric flask. Rinse the receiving flask with a few small portions of distilled water, adding the washings to the volumetric flask. Fill to the mark with distilled water. Stopper the flask and invert it at least 20 times to ensure that the liquid is thoroughly mixed.

Clean two burets. Rinse one of them with several small portions of the NH_3 solution in the volumetric flask, draining some of the solution through the stopcock. Fill the buret with the NH_3 solution and read and record the level. Rinse the other buret with small portions of the standardized HCl solution from the stock supply, and then fill that buret with that solution. Record the level.

Draw about 40 mL of the NH_3 solution into a 250-mL Erlenmeyer flask, and add five drops of bromcresol green indicator. Titrate with HCl to the end point, where the indicator changes from blue to yellow. The actual end point is green and can be reached by back titrating as necessary with NH_3. Record the final levels in the NH_3 and HCl burets.

Repeat the titration once or twice with 40-mL portions of the NH_3 solution. The NH_3-HCl volume ratios should check within 1% if all goes well.

The titrated solutions may be poured down the sink. The liquid remaining in the distilling flask should be discarded in the waste crock.

E. Alternative Procedure for Preparation of $Co_x(NH_3)_yCl_z$ (Optional)

There are several possible $Co_x(NH_3)_yCl_z$ compounds. The following procedure will allow you to make a different one from that produced in Part A of this experiment.

Using a piece of weighing paper and a top-loading balance, weigh out 4.0 ± 0.1 g of NH_4Cl, ammonium chloride. Pour the NH_4Cl into a 250-ml beaker, and **in the hood,** slowly add 25 mL 15 M NH_3. (**C A U T I O N:** *This is a concentrated reagent, with a strong NH_3 odor.*) Cover the beaker with a watch glass. To the solution add 8.0 ± 0.2 g of $CoCl_2 \cdot 6 H_2O$, cobalt(II) chloride hexahydrate while stirring the mixture with a stirring rod. Continue stirring until a brown slurry is produced. Cover the mixture.

Measure out 6 mL of 30% H_2O_2, hydrogen peroxide. (**C A U T I O N:** *This reagent can cause severe skin burns; use gloves while handling.*) **In the hood,** add the H_2O_2 to the mixture in 1-mL portions while stirring. There will be some effervescence as oxygen gas is evolved. Tip the beaker to bring any brown material on the walls of the beaker into contact with the H_2O_2 solution. **In the hood,** measure out 25 mL 12 M HCl. (**C A U T I O N:** *This is a concentrated reagent with a choking odor.*) While stirring the mixture in the beaker, slowly add the HCl. A white smoke will form, due to the NH_4Cl dust produced when the HCl vapor reacts with the NH_3 vapor above the mixture. Cover the beaker.

At your lab bench heat the mixture in the beaker to 80 to 85°C for about 30 minutes. Control the temperature by judicious use of your Bunsen burner.

While the mixture is being heated, prepare two fritted glass crucibles, according to the directions in the 4th paragraph of part A of this experiment.

Remove heat from the beaker after the 30 minutes are up, and let the beaker cool to room temperature. Set up a Buchner funnel, and, with suction, filter the mixture through the filter paper in the funnel. Discard the filtrate in the waste container.

Scrape the solid from the filter paper into the 250-mL beaker. Add 60 mL 6 M NH_3, and heat gently to dissolve the solid. Cool to room temperature. **In the hood,** pour 60 mL of 12 M HCl into a second beaker. Slowly pour the cobalt-containing solution into the HCl solution. Cover the beaker with a watch glass and take it to your lab bench.

Heat the solution to 80 to 85°C for 30 minutes, as before. Cool to room temperature, and filter the mixture with suction, using the Buchner funnel. Your product will be the crystals in the funnel. Turn off the suction after two minutes.

Pour 10 mL of cold 95% ethanol on to the crystals in the funnel, wait 10 seconds, and then turn on the suction, to pull through the ethanol, which should carry with it most of the water and HCl remaining on the crystals. Wash again with a second 10-mL portion of cold ethanol. Carry out two more washings, this time with 10-mL portions of cold acetone. Draw air through the crystals for five minutes.

Weigh a 100-mL beaker to ± 0.1 g. Transfer the crystals from the paper to the beaker. Weigh the beaker and its contents. Cover the beaker and put it in your lab drawer for use in the next lab period. Discard the liquid in the suction flask into the waste container or as otherwise directed by your instructor.

Data and Calculations: Synthesis and Analysis of a Coordination Compound

A. Preparation of $Co_x(NH_3)_yCl_z$

Mass of $CoCl_2 \cdot 6 H_2O$ _____ g

Mass of 100-mL beaker _____ g

Approximate mass of product (from B) _____ g

B. Determination of Chloride Ion

Mass of 100-mL beaker plus product _____ g

Mass after first transfer _____ g	Mass of sample I	_____ g
Mass after second transfer _____ g	Mass of sample II	_____ g

Mass of first filter crucible _____ g

Mass of second filter crucible _____ g

Mass of first crucible plus AgCl _____ g	Mass of AgCl I	_____ g
Mass of second crucible plus AgCl _____ g	Mass of AgCl II	_____ g

C. Determination of Cobalt Ion

Mass of 100-mL beaker plus product _____ g

Mass after first transfer _____ g	Mass of sample I	_____ g
Mass after second transfer _____ g	Mass of sample II	_____ g

Absorbance of first solution _____

Molarity of Co ion _____ M

Absorbance of second solution _____

Molarity of Co ion _____ M

(continued on following page)

(continued)

D. Determination of Ammonia

Mass of 100-mL beaker plus product   _____ g

Mass after transfer   _____ g   *2nd trial*

Initial reading NH_3 buret   _____ mL   _____ mL

Final reading NH_3 buret   _____ mL   _____ mL

Initial reading HCl buret   _____ mL   _____ mL

Final reading HCl buret   _____ mL   _____ mL

Molarity of standardized HCl   _____ M

E. Calculation of Chloride Content

	Sample I	*Sample II*
Mass of AgCl	_____ g	_____ g
Mass of Cl^- in AgCl	_____ g	_____ g
Mass of sample	_____ g	_____ g
Mass of Cl^- per 100-g sample	_____ g	_____ g
Moles Cl^- per 100-g sample	_____	_____

F. Calculation of Cobalt Content

	Sample I	*Sample II*
Molarity of cobalt ion	_____ M	_____ M
Moles cobalt in 25 mL = moles cobalt in sample	_____	_____
Mass of sample	_____ g	_____ g
Moles cobalt per 100-g sample	_____	_____

(continued on following page)

G. Calculation of Ammonia Content

Mass of sample _____ g   Molarity of HCl _____ M

	Trial I	*Trial II*
Volume NH_3 used	_____ mL	_____ mL
Volume HCl used	_____ mL	_____ mL
Moles HCl = moles NH_3	_____	_____
Moles NH_3 per 250-mL NH_3 solution = moles NH_3 in sample	_____	_____
Moles NH_3 per 100-g sample	_____	_____

H. Determination of the Formula of $Co_x(NH_3)_yCl_z$

For a 100-g sample: Moles Cl^- _____   Moles Co ion _____   Moles NH_3 _____

Whole number ratio: $z =$ _____ $x =$ _____ $y =$ _____

Formula of compound:

Underline the procedure you used: Part A   Part E

Advance Study Assignment: Synthesis and Analysis of a Coordination Compound

A student prepared 3.8 g of $Co_x(NH_3)_yCl_z$. She then analyzed the compound by the procedure in this experiment.

A.   In the gravimetric determination of chloride, she weighed out 0.3011 g of the compound. The following data were obtained:

Mass of crucible plus AgCl   18.7137 g

Mass of crucible   18.2316 g

Mass of AgCl   ——————— g   MM AgCl = 143.34 g

Mass of Cl^- in AgCl   ——————— g   MM Cl^- = 35.45 g

Moles Cl^- in sample   ———————

Moles Cl^- in 100-g sample   ———————

B.   In the colorimetric determination of cobalt, she used a sample weighing 0.4802 g. The molarity of cobalt ion in the solution from the volumetric flask was 0.072 M.

Moles cobalt ion in 25-mL solution = moles cobalt in sample   ———————

Moles cobalt per 100-g sample   ———————

C.   In the volumetric determination of ammonia, the sample weighed 1.0014 g. In the titration, 0.1000 M HCl was used. She found that 40.00 mL of the NH_3 solution required 30.13 mL of the HCl to reach the end point.

No. moles HCl used   ——————— = no. moles NH_3 in 40-mL NH_3 solution

No. moles NH_3 in 250-mL NH_3 ——————— = no. moles NH_3 in sample

No. moles NH_3 per 100-g sample   ———————

(continued on following page)

(continued)

D. Calculation of the formula of $Co_x(NH_3)_yCl_z$:

In 100 g of sample, no. moles Co = _____

no. moles NH_3 = _____

no. moles Cl^- = _____

Dividing by smallest number, no. moles Co ion _____

no. moles NH_3 _____

no. moles Cl^- _____

Formula of compound _____

(This is probably not the formula of the compound you will be making, but it could have been.)

DETERMINATION OF IRON BY REACTION WITH PERMANGANATE—A REDOX TITRATION

Potassium permanganate, $KMnO_4$, is widely used as an oxidizing agent in volumetric analysis. In acid solution, MnO_4^- ion undergoes reduction to Mn^{2+} as shown in the following equation:

$$8\,H^+(aq) + MnO_4^-(aq) + 5\,e^- \rightarrow Mn^{2+}(aq) + 4\,H_2O$$

Since the MnO_4^- ion is violet and the Mn^{2+} ion is nearly colorless, the end point in titrations using $KMnO_4$ as the titrant can be taken as the first permanent pink color that appears in the solution.

$KMnO_4$ will be employed in this experiment to determine the percentage of iron in an unknown containing iron(II) ammonium sulfate, $Fe(NH_4)_2(SO_4)_2 \cdot 6\,H_2O$. The titration, which involves the oxidation of Fe^{2+} ion to Fe^{3+} by permanganate ion, is carried out in sulfuric acid solution to prevent the air-oxidation of Fe^{2+}. The end point of the titration is sharpened markedly if phosphoric acid is present, because the Fe^{3+} ion produced in the titration forms an essentially colorless complex with the acid.

The number of moles of potassium permanganate used in the titration is equal to the product of the molarity of the $KMnO_4$ and the volume used. The number of moles of iron present in the sample is obtained from the balanced equation for the reaction and the amount of MnO_4^- ion reacted. The percentage by weight of iron in the solid sample follows directly.

Experimental Procedure

Obtain from the stockroom a buret and an unknown iron(II) sample.

Weigh out accurately on the analytical balance three samples of about 1.0 g of your unknown into clean 250-mL Erlenmeyer flasks.

Clean your buret thoroughly. Draw about 100 mL of the standard $KMnO_4$ solution from the carboy in the laboratory. Rinse the buret with a few milliliters of the $KMnO_4$ three times. Drain and then fill the buret with the $KMnO_4$ solution.

Prepare 150 mL of 1 M H_2SO_4 by pouring 25 mL of 6 M H_2SO_4 into 125 mL of H_2O, while stirring. Add 50 mL of this 1 M H_2SO_4 to *one* of the iron samples. The sample should dissolve completely. Without delay, titrate this iron solution with the $KMnO_4$ solution. When a light-yellow color develops in the iron solution during the titration, add 3 mL of 85% H_3PO_4. **CAUTION:** *Caustic reagent.*

Continue the titration until you obtain the first pink color that persists for 15 to 30 seconds. Repeat the titration with the other two samples.

(Optional) The $KMnO_4$ solution can be standardized by the method you used in this experiment. $Fe(NH_4)_2(SO_4)_2 \cdot 6\,H_2O$ is a primary standard with a molar mass equal to

392.2 g. For standardization, use 0.7 ± 0.05 g samples of the primary standard, weighed to 0.0001 g. Draw 100 mL of the stock $KMnO_4$ solution and titrate the samples as you did the unknown.

Dispose of your titrated solutions as directed by your laboratory supervisor.

Data and Calculations: Determination of Iron by Permanganate Titration

Mass of sample tube plus contents ————————— g

Mass after removing Sample I ————————— g

Mass after removing Sample II ————————— g

Mass after removing Sample III ————————— g

Molarity of standard $KMnO_4$ solution ————————— M

Sample	*I*	*II*	*III*
Initial buret reading	————————— mL	————————— mL	————————— mL
Final buret reading	————————— mL	————————— mL	————————— mL
Volume $KMnO_4$ required	————————— mL	————————— mL	————————— mL
Moles $KMnO_4$ required	—————————	—————————	—————————
Moles Fe^{2+} in sample	—————————	—————————	—————————
Mass of Fe in sample	————————— g	————————— g	————————— g
Mass of sample	————————— g	————————— g	————————— g
Percentage of Fe in sample	————————— %	————————— %	————————— %

Unknown no. —————————

Standardization of $KMnO_4$ (Optional)

Sample	*IV*	*V*	*VI*
Mass of primary standard	————————— g	————————— g	————————— g
Moles Fe in standard	—————————	—————————	—————————
Moles $KMnO_4$ required	—————————	—————————	—————————

(continued on following page)

(continued)

Initial buret reading	_____ mL	_____ mL	_____ mL
Final buret reading	_____ mL	_____ mL	_____ mL
Volume KMnO$_4$ used	_____ mL	_____ mL	_____ mL
Molarity KMnO$_4$	_____ M	_____ M	_____ M
Average molarity	_____ M		

Advance Study Assignment: Determination of Iron by Permanganate Titration

1. Write the balanced net ionic equation for the reaction between MnO_4^- ion and Fe^{2+} ion in acid solution.

2. How many moles of Fe^{2+} ion can be oxidized by 1.4×10^{-2} moles MnO_4^- ion in the reaction in Question 1?

_____ moles

3. A solid sample containing some Fe^{2+} ion weighs 1.062 g. It requires 24.12 mL 0.01562 M $KMnO_4$ to titrate the Fe^{2+} in the dissolved sample to a pink end point.

a. How many moles MnO_4^- ion are required?

_____ moles

b. How many moles Fe^{2+} are there in the sample?

_____ moles

c. How many grams of iron are there in the sample?

_____ g

d. What is the percentage of Fe in the sample?

_____ %

4. What is the percentage of Fe in iron(II) ammonium sulfate hexahydrate, $Fe(NH_4)_2(SO_4)_2 \cdot 6 \, H_2O$?

_____ %

DETERMINATION OF AN EQUIVALENT MASS BY ELECTROLYSIS

In Experiment 22 we determined the equivalent mass of an acid by titration of a known mass of the acid with a standardized solution of NaOH. There we defined an equivalent mass of acid to be the amount of acid in grams that could furnish one mole of H^+ ion.

Experimentally we find that the equivalent mass of an element can be related in a fundamental way to the chemical effects observed in that phenomenon known as *electrolysis*. As you know, some liquids, because they contain ions, will conduct an electric current. If the two terminals on a storage battery, or any other source of D.C. voltage, are connected through metal electrodes to a conducting liquid, an electric current will pass through the liquid and chemical reactions will occur at the two metal electrodes; in this process electrolysis is said to occur, and the liquid is said to be electrolyzed.

At the electrode connected to the *negative* pole of the battery, a *reduction* reaction will invariably be observed. In this reaction electrons will usually be accepted by one of the species present in the liquid, which, in the experiment we shall be doing, will be an aqueous solution. The species reduced will ordinarily be a metallic cation or the H^+ ion or possibly water itself; the reaction that is actually observed will be the one that occurs with the least expenditure of electrical energy, and will depend on the composition of the solution. In the electrolysis cell we shall study, the reduction reaction of interest will occur in a slightly acidic medium; hydrogen gas will be produced by the reduction of hydrogen ion:

$$2\,H^+(aq) + 2\,e^- \rightarrow H_2(g) \tag{1}$$

In this reduction reaction, which will occur at the negative pole, or *cathode,* of the cell, for every H^+ ion reduced *one* electron will be required, and for every molecule of H_2 formed, *two* electrons will be needed.

Ordinarily in chemistry we deal not with individual ions or molecules but rather with moles of substances. In terms of moles, we can say that, by Equation 1,

> The reduction of one mole of H^+ ion requires one mole of electrons
> The production of one mole of $H_2(g)$ requires two moles of electrons

A mole of electrons is a fundamental amount of electricity in the same way that a mole of pure substance is a fundamental unit of matter, at least from a chemical point of view. A mole of electrons is called a faraday, after Michael Faraday, who discovered the basic laws of electrolysis. The amount of a species which will react with a *mole* of *electrons,* or *one faraday,* is equal to the *equivalent mass* of that species. Since one faraday will reduce one mole of H^+ ion, we say that the equivalent mass of hydrogen is 1.008 grams, equal to the mass of one mole of H^+ ion (or 1/2 mole of $H_2(g)$). To form one mole of $H_2(g)$ one would have to pass two faradays through the electrolysis cell.

In the electrolysis experiment we will perform we will measure the volume of hydrogen gas produced under known conditions of temperature and pressure. By using the Ideal Gas Law we will be able to calculate how many moles of H_2 were formed, and hence how many faradays of electricity passed through the cell.

At the positive pole of an electrolysis cell (the metal electrode that is connected to the + terminal of the battery), an *oxidation* reaction will occur, in which some species will give up electrons. This reaction, which takes place at the *anode* in the cell, may involve again an ionic or neutral species in the solution or the metallic electrode itself. In the cell that you will be studying, the pertinent oxidation reaction will be that in which a metal under study will participate:

$$M(s) \rightarrow M^{n+}(aq) + ne^- \tag{2}$$

During the course of the electrolysis the atoms in the metal electrode will be converted to metallic cations and will go into the solution. The mass of the metal electrode will decrease, depending on the amount of electricity passing through the cell and the nature of the metal. To oxidize one mole, or one molar mass, of the metal, it would take n faradays, where n is the charge on the cation that is formed. By definition, one faraday of electricity would cause one equivalent mass, EM, of metal to go into solution. The molar mass, MM, and the equivalent mass of the metal are related by the equation:

$$MM = EM \times n \tag{3}$$

In an electrolysis experiment, since n is not determined independently, it is not possible to find the molar mass of a metal. It is possible, however, to find equivalent masses of many metals, and that will be our main purpose.

The general method we will use is implied by the discussion. We will oxidize a sample of an unknown metal at the positive pole of an electrolysis cell, weighing the metal before and after the electrolysis and so determining its loss in mass. We will use the same amount of electricity, the same number of electrons, to reduce hydrogen ion at the negative pole of the electrolysis cell. From the volume of H_2 gas that is produced under known conditions we can calculate the number of moles of H_2 formed, and hence the number of faradays that passed through the cell. The equivalent mass of the metal is then calculated as the amount of metal that would be oxidized if one faraday were used. In an optional part of the experiment, your instructor may tell you the nature of the metal you used. Using Equation 3, it will be possible to determine the charge on the metallic cations that were produced during electrolysis.

Experimental Procedure

Obtain from the stockroom a buret and a sample of metal unknown. Lightly sand the metal to clean it. Rinse the metal with water and then in acetone. Let the acetone evaporate. When the sample is dry, weigh it on the analytical balance to 0.0001 g.

Set up the electrolysis apparatus as indicated in Figure 29.1. There should be about 100 mL 0.5 M $HC_2H_3O_2$ in 0.5 M Na_2SO_4 in the beaker with the gas buret. This will serve as the conducting solution. Immerse the end of the buret in the solution and attach a length of rubber tubing to its upper end. Open the stopcock on the buret and, with suction, carefully draw the acid up to the top of the graduations. Close the stopcock. Insert the bare coiled end of the heavy copper wire up into the end of the buret; all but the coil end of the wire should be covered with watertight insulation. Check the solution level after a few minutes to make sure the stopcock does not leak. Record the level.

The metal unknown will serve as the anode in the electrolysis cell. Connect the metal to the + pole of the power source with an alligator clip and immerse the metal but not the clip in the conducting solution. The copper electrode will be the cathode in the cell. Connect that electrode to the − pole of the power source. Hydrogen gas should immediately begin to bubble from the copper cathode. Collect the gas until about 50 mL have been produced. At that point, stop the electrolysis by disconnecting the copper electrode from the power source.

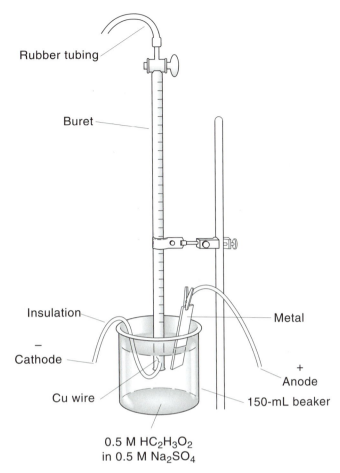

Rubber tubing

Buret

Insulation

–
Cathode

Cu wire

Metal

+
Anode

150-mL beaker

0.5 M HC$_2$H$_3$O$_2$
in 0.5 M Na$_2$SO$_4$

FIGURE 29.1

Record the level of the liquid in the buret. Measure and record the temperature and the barometric pressure in the laboratory. (In some cases a cloudiness may develop in the solution during the electrolysis. This is caused by the formation of a metal hydroxide, and will have no adverse effect on the experiment.)

Raise the buret, and discard the conducting solution in the beaker in the waste crock. Rinse the beaker with water, and pour in 100 mL of fresh conducting solution. Repeat the electrolysis, generating about 50 mL of H$_2$ and recording the initial and final liquid levels in the buret. Take the alligator clip off the metal anode and wash the anode with 0.1 M HC$_2$H$_3$O$_2$, acetic acid. Rub off any loose adhering coating with your fingers, and then rinse off your hands. Rinse the electrode in water and then in acetone. Let the acetone evaporate. Weigh the dry metal electrode to the nearest 0.0001 g.

When you are finished with the experiment, return the metal electrode to the stockroom and discard the conducting solution in the waste crock unless directed otherwise by your instructor.

Data and Calculations: Determination of an Equivalent Mass
by Electrolysis

Mass of metal anode _____ g

Mass of anode after electrolysis _____ g

Initial buret reading _____ mL

Buret reading after first electrolysis _____ mL

Buret reading after refilling _____ mL

Buret reading after second electrolysis _____ mL

Barometric pressure _____ mm Hg

Temperature t _____ °C

Vapor pressure of H_2O at t _____ mm Hg

Total volume of H_2 produced, V _____ mL

Temperature T _____ K

Pressure exerted by dry H_2: $P = P_{bar} - VP_{H_2O}$
(ignore any pressure effect due to liquid levels in buret) _____ mm Hg

No. moles H_2 produced, n
(use Ideal Gas Law, $PV = nRT$) _____ moles

No. of faradays passed (no. of moles of electrons) _____

Loss in mass by anode _____ g

Equivalent mass of metal $\left(EM = \dfrac{\text{no. g lost}}{\text{no. faradays passed}} \right)$ _____ g

Unknown metal number _____

Optional: Nature of metal _____ MM _____ g

Charge n on cation _____ (Eq. 3)

Advance Study Assignment: Determination of an Equivalent Mass by Electrolysis

1. In an electrolysis cell similar to the one employed in this experiment, a student observed that his unknown metal anode lost 0.208 g while a total volume of 96.30 mL of H_2 was being produced. The temperature in the laboratory was 25°C and the barometric pressure was 748 mm Hg. At 25°C the vapor pressure of water is 23.8 mm Hg. To find the equivalent mass of his metal, he filled in the blanks below. Fill in the blanks as he did.

$P_{H_2} = P_{bar} - VP_{H_2O} =$ _____ mm Hg = _____ atm

$V_{H_2} =$ _____ mL = _____ L

$T =$ _____ K

$n_{H_2} =$ _____ moles $n_{H_2} = \dfrac{PV}{RT}$ (where $P = P_{H_2}$)

1 mole H_2 requires passage of _____ faradays

No. of faradays passed = _____

Loss of mass of metal anode = _____ g

No. grams of metal lost per faraday passed $= \dfrac{\text{no. grams lost}}{\text{no. faradays passed}} =$ _____ g = EM

The student was told that his metal anode was made of iron.

MM Fe = _____ g. The charge n on the Fe ion is therefore _____. (See Eq. 3.)

2. In ordinary units, the faraday is equal to 96,480 coulombs. A coulomb is the amount of electricity passed when a current of one ampere flows for one second. Given the charge on an electron, 1.6022×10^{-19} coulombs, calculate a value for Avogadro's number.

VOLTAIC CELL MEASUREMENTS

Many chemical reactions can be classified as oxidation-reduction reactions, because they involve the oxidation of one species and the reduction of another. Such reactions can conveniently be considered as the result of two half-reactions, one of oxidation and the other reduction. In the case of the oxidation-reduction reaction

$$Zn(s) + Pb^{2+}(aq) \rightarrow Zn^{2+}(aq) + Pb(s)$$

that would occur if a piece of metallic zinc were put into a solution of lead nitrate, the two reactions would be

$$Zn(s) \rightarrow Zn^{2+}(aq) + 2\ e^- \quad \text{oxidation}$$

$$2\ e^- + Pb^{2+}(aq) \rightarrow Pb(s) \quad \text{reduction}$$

The tendency for an oxidation-reduction reaction to occur can be measured if the two reactions are made to occur in separate regions connected by a barrier that is porous to ion movement. An apparatus called a voltaic cell in which this reaction might be carried out under this condition is shown in Figure 30.1.

If we connect a voltmeter between the two electrodes, we will find that there is a voltage, or potential, between them. The magnitude of the potential is a direct measure of the driving force or thermodynamic tendency of the spontaneous oxidation-reduction reaction to occur.

If we study several oxidation-reduction reactions, we find that the voltage of each associated voltaic cell can be considered to be the sum of a potential for the oxidation reaction and

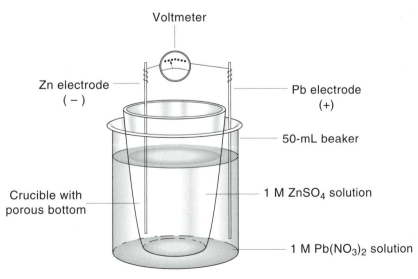

Voltmeter

Zn electrode
(−)

Pb electrode
(+)

50-mL beaker

1 M ZnSO₄ solution

Crucible with
porous bottom

1 M Pb(NO₃)₂ solution

FIGURE 30.1

a potential for the reduction reaction. In the Zn,Zn^{2+} ‖ Pb^{2+},Pb cell we have been discussing, for example,

$$E_{cell} = E_{Zn,Zn^{2+}\text{oxidation reaction}} + E_{Pb^{2+},Pb\text{ reduction reaction}} \tag{1}$$

By convention, the negative electrode in a voltaic cell is taken to be the one from which electrons are emitted (i.e., where oxidation occurs). The negative electrode is the one that is connected to the minus pole of the voltmeter when the voltage is measured.

Since any cell potential is the sum of two electrode potentials, it is not possible, by measuring cell potentials, to determine individual absolute electrode potentials. However, if a value of potential is arbitrarily assigned to one electrode reaction, other electrode potentials can be given definite values, based on the assigned value. The usual procedure is to assign a value of 0.0000 volts to the standard potential for the electrode reaction

$$2\,H^+(aq) + 2\,e^- \rightarrow H_2(g); \quad E^0_{H^+,H_2 \text{ red}} = 0.0000\ V$$

For the Zn,Zn^{2+} ‖ H$^+$,H$_2$ cell, the measured potential is 0.76 V, and the zinc electrode is negative. Zinc metal is therefore oxidized, and the cell reaction must be

$$Zn(s) + 2\,H^+(aq) \rightarrow Zn^{2+}(aq) + H_2(g); \quad E^0_{cell} = 0.76\ V$$

Given this information, one can readily find the potential for the oxidation of Zn to Zn^{2+}.

$$E^0_{cell} = E^0_{Zn,Zn^{2+}\text{ oxid}} + E^0_{H^+,H_2 \text{ red}}$$

$$0.76\ V = E^0_{Zn,Zn^{2+}\text{ oxid}} + 0.00\ V; \quad E^0_{Zn,Zn^{2+}\text{ oxid}} = +0.76\ V$$

If the potential for a half-reaction is known, the potential for the reverse reaction can be obtained by changing the sign. For example:

$$\textit{if} \quad E^0_{Zn,Zn^{2+}\text{ oxid}} = +0.76\ V, \quad \textit{then} \quad E^0_{Zn^{2+},Zn \text{ red}} = -0.76\ V$$

$$\textit{if} \quad E^0_{Pb^{2+},Pb \text{ red}} = +Y\ V, \quad \textit{then} \quad E^0_{Pb,Pb^{2+}\text{ oxid}} = -Y\ V$$

In the first part of this experiment you will measure the voltages of several different cells. By arbitrarily assigning the potential of a particular half-reaction to be 0.00 V, you will be able to calculate the potentials corresponding to all of the various half-reactions that occurred in your cells.

In our discussion so far we have not considered the possible effects of such system variables as temperature, potential at the liquid-liquid junction, size of metal electrodes, and concentrations of solute species. Although temperature and liquid junctions do have a definite effect on cell potentials, taking account of their influence involves rather complex thermodynamic concepts and is usually not of concern in an elementary course. The size of a metal electrode has no appreciable effect on electrode potential, although it does relate directly to the capacity of the cell to produce useful electrical energy. In this experiment we will operate the cells so that they deliver essentially no energy but exert their maximum potentials.

The effect of solute ion concentrations is important and can be described relatively easily. For the cell reaction at 25°C:

$$aA(s) + bB^+(aq) \rightarrow cC(s) + dD^{2+}(aq)$$

$$E_{cell} = E^0_{cell} - \frac{0.0592}{n} \log \frac{[D^{2+}]^d}{[B^+]^b} \tag{2}$$

where E^0_{cell} is a constant for a given reaction and is called the standard cell potential, and n is the number of electrons in either electrode reaction.

By Equation 2 you can see that the measured cell potential, E_{cell}, will equal the standard cell potential if the molarities of D^{2+} and B^+ are both unity, or, if d equals b, if the molarities are simply equal to each other. We will carry out experiments under such conditions that the cell potentials you observe will be very close to the standard potentials given in the tables in your chemistry text.

Considering the Cu,Cu^{2+} ∥ Ag^+,Ag cell as a specific example, the observed cell reaction would be

$$Cu(s) + 2\ Ag^+(aq) \rightarrow Cu^{2+}(aq) + 2\ Ag(s)$$

For this cell, Equation 2 takes the form

$$E_{cell} = E^0_{cell} - \frac{0.0592}{2} \log \frac{[Cu^{2+}]}{[Ag^+]^2} \tag{3}$$

In the equation n is 2 because in the cell reaction two electrons are transferred in each of the two half-reactions. E^0 would be the cell potential when the copper and silver salt solutions are both 1 M, since then the logarithm term is equal to zero.

If we decrease the Cu^{2+} concentration, keeping that of Ag^+ at 1 M, the potential of the cell will go up by about 0.03 volts for every factor of ten by which we decrease $[Cu^{2+}]$. Ordinarily it is not convenient to change concentrations of an ion by several orders of magnitude, so in general, concentration effects in cells are relatively small. However, if we should add a complexing or precipitating species to the copper salt solution, the value of $[Cu^{2+}]$ would drop drastically, and the voltage change would be appreciable. In the experiment we will illustrate this effect by using NH_3 to complex the Cu^{2+}. Using Equation 3, we can actually calculate $[Cu^{2+}]$ in the solution of its complex ion. It is very low.

In an analogous experiment we will determine the solubility product of AgCl. In this case we will surround the Ag electrode in a Cu,Cu^{2+} ∥ Ag^+, Ag cell with a solution of known Cl^- ion concentration that is saturated with AgCl. From the measured cell potential, we can use Equation 3 to calculate the very small value of $[Ag^+]$ in the chloride-containing solution.

In the two experiments in which we change the cation concentrations, first Cu^{2+} and then Ag^+, we end up with systems in equilibrium. In the first case Cu^{2+} ion is in equilibrium with $Cu(NH_3)_4^{2+}$ and NH_3. Using the cell potential, we can calculate $[Cu^{2+}]$. From the way we made up the system we can calculate $[Cu(NH_3)_4^{2+}]$ and $[NH_3]$. If we wish to do so, we can use these data to find the dissociation constant for the $Cu(NH_3)_4^{2+}$ ion. Many equilibrium constants are found by experiments of this sort.

In the last experiment we have an equilibrium system containing Ag^+, Cl^-, and AgCl(s). Here we can use the cell potential to find $[Ag^+]$. There isn't much Ag^+ present, but there is a tiny bit, and the cell potential lets us find it. We know $[Cl^-]$ from the way we made up the mixture in the crucible. From these data we can easily calculate K_{sp} for AgCl:

$$AgCl(s) \rightleftarrows Ag^+(aq) + Cl^-(aq) \qquad K_{sp} = [Ag^+]\,[Cl^-] \tag{4}$$

Experimental Procedure

You may work in pairs in this experiment.

Cell Potentials

In this experiment you will be working with these seven electrode systems:

Ag^+, $Ag(s)$	$Br_2(l)$, Br^-,Pt
Cu^{2+}, $Cu(s)$	$Cl_2(g$, 1 atm$)$, Cl^-,Pt
Fe^{3+}, Fe^{2+},Pt	$I_2(s)$,I^-, Pt
Zn^{2+}, $Zn(s)$	

Your purpose will be to measure enough voltaic cell potentials to allow you to determine the electrode potentials of each electrode by comparing it with an arbitrarily chosen electrode potential.

Using the apparatus shown in Figure 30.1, set up a voltaic cell involving any two of the electrodes in the list. About 10 mL of each solution should be enough for making the cell. The solute ion concentrations may be assumed to be one molar and all other species may be assumed to be at unit activity, so that the potentials of the cells you set up will be essentially the standard potentials. Measure the cell potential and record it along with which electrode has negative polarity.

In a similar manner set up and measure the cell voltage and polarities of other cells, sufficient in number to include all of the electrode systems on the list at least once. Do not combine the silver electrode system with any of the halogen electrode systems because a precipitate will form; any other combinations may be used. The data from this part of the experiment should be entered in the first three columns of the table in Part A.1 of your report.

B. Effect of Concentration on Cell Potentials

1. COMPLEX ION FORMATION. Set up the Cu,Cu^{2+} ‖ Ag^+,Ag cell, using 10 mL of the $CuSO_4$ solution in the crucible and 10 mL of $AgNO_3$ in the beaker. Measure the potential of the cell. While the potential is being measured, add 10 mL of 6 M NH_3 to the $CuSO_4$ solution, stirring carefully with your stirring rod. Measure the potential when it becomes steady.

2. DETERMINATION OF THE SOLUBILITY PRODUCT OF AgCl. Remove the crucible from the cell you have just studied and discard the $Cu(NH_3)_4^{2+}$. Clean the crucible by drawing a little 6 M NH_3 through it, using the adapter and suction flask. Then draw some distilled water through it. Reassemble the Cu-Ag cell, this time using the beaker for the Cu-$CuSO_4$ electrode system. Put 10 mL 1 M KCl in the crucible; immerse the Ag electrode in that solution. Add a drop of $AgNO_3$ solution to form a little AgCl, so that an equilibrium between Ag^+ and Cl^- can be established. Measure the potential of this cell, noting which electrode is negative. In this case [Ag^+] will be very low, which will decrease the potential of the cell to such an extent that its polarity may change from that observed previously.

DISPOSAL OF REAGENTS. As you finish each voltage measurement, pour the two solutions into a beaker. At the end of the experiment, pour the contents of the beaker into the waste crock, unless directed otherwise by your instructor.

Data and Calculations: Voltaic Cell Measurements

A.1. Cell Potentials

Electrode Systems Used in Cell	Cell Potential, E^0_{cell} (volts)	Negative Electrode	Oxidation Reaction	$E^0_{oxidation}$ in volts	Reduction Reaction	$E^0_{reduction}$ in volts
1.						
2.						
3.						
4.						
5.						
6.						
7.						

Calculations

A. Noting that oxidation occurs at the *negative* pole in a cell, write the oxidation reaction in each of the cells. The other electrode system must undergo reduction; write the reduction reaction that occurs in each cell.

B. Assume that $E^0_{Ag^+, Ag} = 0.00$ volts (whether in reduction or oxidation). Enter that value in the table for all of the silver electrode systems you used in your cells. Since $E^0_{cell} = E^0_{oxidation} + E^0_{reduction}$, you can calculate E^0 values for all the electrode systems in which the Ag,Ag$^+$ system was involved. Enter those values in the table.

C. Using the values and relations in (B) and taking advantage of the fact that for any given electrode system, $E^0_{oxidation} = -E^0_{reduction}$, complete the table of E^0 values. The best way to do this is to use one of the E^0 values you found in (B) in another cell with that electrode system. That potential, along with E^0_{cell}, will allow you to find the potential of the other electrode. Continue this process with other cells until all the electrode potentials have been determined.

(continued on following page)

(continued)

A.2. Table of Electrode Potentials

In Table A.1, you should have a value for E_{red}^0 or E_{oxid}^0 for each of the electrode systems you have studied. Remembering that for any electrode system, $E_{red}^0 = -E_{oxid}^0$, you can find the value for E_{red}^0 for each system. List those potentials in the left column of the table below in order of decreasing value.

$E_{reduction}^0$ ($E_{Ag^+,Ag}^0 = 0.00$ V)	Electrode Reaction in Reduction	$E_{reduction}^0$ ($E_{H^+,H_2}^0 = 0.00$ V)
_____	_____	_____
_____	_____	_____
_____	_____	_____
_____	_____	_____
_____	_____	_____
_____	_____	_____
_____	_____	_____

The electrode potentials you have determined are based on $E_{Ag^+,Ag}^0 = 0.00$ V. The usual assumption is that $E_{H^+,H_2}^0 = 0.00$ volts, under which conditions $E_{Ag^+,Ag\ red}^0 = 0.80$ V. Convert from one base to the other by adding 0.80 volts to each of the electrode potentials, and enter these values in the third column of the table.

Why are the values of E_{red}^0 on the two bases related to each other in such a simple way?

B. Effect of Concentration on Cell Potentials

1. Complex ion formation:

Potential, E_{cell}^0, before addition of 6 M NH_3 _____ V

Potential, E_{cell}, after $Cu(NH_3)_4^{2+}$ formed   _____ V

Given Equation 3

$$E_{cell} = E_{cell}^0 - \frac{0.0592}{2} \log \frac{[Cu^{2+}]}{[Ag^+]^2} \tag{3}$$

(continued on following page)

(continued)

calculate the residual concentration of free Cu^{2+} ion in equilibrium with $Cu(NH_3)_4^{2+}$ in the solution in the crucible. Take $[Ag^+]$ to be 1 M.

$[Cu^{2+}] = $ _____ M

2. Solubility product of AgCl:

Potential, E^0_{cell}, of the $Cu,Cu^{2+} \parallel Ag^+,Ag$ cell (from B.1)

_____ V Negative electrode _____

Potential, E_{cell}, with 1 M KCl present   _____ V Negative electrode _____

Using Equation 3, calculate $[Ag^+]$ in the cell, where it is in equilibrium with 1 M Cl^- ion. (E_{cell} in Equation 3 is the *negative* of the measured value if the polarity is not the same as in the standard cell.) Take $[Cu^{2+}]$ to be 1 M.

$[Ag^+] = $ _____ M

Since Ag^+ and Cl^- in the crucible are in equilibrium with AgCl, we can find K_{sp} for AgCl from the concentrations of Ag^+ and Cl^-, which we now know. Formulate the expression for K_{sp} for AgCl, and determine its value.

$K_{sp} = $ _____

Advance Study Assignment: Voltaic Cell Measurements

1. A student measures the potential of a cell made up with 1 M $CuSO_4$ in one solution and 1 M $AgNO_3$ in the other. There is a Cu electrode in the $CuSO_4$ and an Ag electrode in the $AgNO_3$, and the cell is set up as in Figure 30.1. She finds that the potential, or voltage, of the cell, E^0_{cell}, is 0.45 V, and that the Cu electrode is negative.

   a. At which electrode is oxidation occurring?   ————————————

   b. Write the equation for the oxidation reaction.

   c. Write the equation for the reduction reaction.

   d. If the potential of the silver, silver ion electrode, $E^0_{Ag^+,Ag}$ is taken to be 0.000 V in oxidation or reduction, what is the value of the potential for the oxidation reaction, $E^0_{Cu,Cu^{2+} \ oxid}$? $E^0_{cell} = E^0_{oxid} + E^0_{red}$.

                                                                    ———————————— volts

   e. If $E^0_{Ag^+,Ag \ red}$ equals 0.80 V, as in standard tables of electrode potentials, what is the value of the potential of the oxidation reaction of copper, $E^0_{Cu,Cu^{2+} \ oxid}$?

                                                                    ———————————— volts

   f. Write the net ionic equation for the spontaneous reaction that occurs in the cell that the student studied.

   g. The student adds 6 M NH_3 to be $CuSO_4$ solution until the Cu^{2+} ion is essentially all converted to $Cu(NH_3)_4^{2+}$ ion. The voltage of the cell, E_{cell}, goes up to 0.92 V and the Cu electrode is still negative. Find the residual concentration of Cu^{2+} ion in the cell. (Use Eq. 3.)

                                                                    ———————————— M

   h. In Part g, $[Cu(NH_3)_4^{2+}]$ is about 0.05 M, and $[NH_3]$ is about 3 M. Given those values and the result in Part 1g for $[Cu^{2+}]$, calculate K for the reaction:

   $$Cu(NH_3)_4^{2+}(aq) \rightleftarrows Cu^{2+}(aq) + 4 \ (NH_3)(aq)$$

                                                                    ————————————

PREPARATION OF COPPER(I) CHLORIDE

Oxidation-reduction reactions are, like precipitation reactions, often used in the preparation of inorganic substances. In this experiment we will employ a series of such reactions to prepare one of the less commonly encountered salts of copper, copper(I) chloride. Most copper compounds contain copper(II), but copper(I) is present in a few slightly soluble or complex copper salts.

The process of synthesis of CuCl we will use begins by dissolving copper metal in nitric acid:

$$Cu(s) + 4\ H^+(aq) + 2\ NO_3^-(aq) \rightarrow Cu^{2+}(aq) + 2\ NO_2(g) + 2\ H_2O \tag{1}$$

The solution obtained is treated with sodium carbonate in excess, which neutralizes the remaining acid with evolution of CO_2 and precipitates Cu(II) as the carbonate:

$$2\ H^+(aq) + CO_3^{2-}(aq) \rightleftharpoons (H_2CO_3)(aq) \rightleftharpoons CO_2(g) + H_2O \tag{2}$$

$$Cu^{2+}(aq) + CO_3^{2-}(aq) \rightleftharpoons CuCO_3(s) \tag{3}$$

The $CuCO_3$ will be purified by filtration and washing and dissolved in hydrochloric acid. Copper metal added to the highly acidic solution then reduces the Cu(II) to Cu(I) and is itself oxidized to Cu(I) in a disproportionation reaction. In the presence of excess chloride, the copper will be present as a $CuCl_4^{3-}$ complex ion. Addition of this solution to water destroys the complex, and white CuCl precipitates.

$$CuCO_3(s) + 2\ H^+(aq) + 4\ Cl^-(aq) \rightarrow CuCl_4^{2-}(aq) + CO_2(g) + H_2O \tag{4}$$

$$CuCl_4^{2-}(aq) + Cu(s) + 4\ Cl^-(aq) \rightarrow 2\ CuCl_4^{3-}(aq) \tag{5}$$

$$CuCl_4^{3-}(aq) \xrightarrow{H_2O} CuCl(s) + 3\ Cl^-(aq) \tag{6}$$

Because CuCl is readily oxidized, due care must be taken to minimize its exposure to air during its preparation and while it is being dried.

In an optional part of the experiment, we prove that the formula of the compound synthesized is CuCl.

Experimental Procedure

WEAR YOUR SAFETY GLASSES WHILE
PERFORMING THIS EXPERIMENT

Obtain a 1-gram sample of copper metal turnings, a Buchner funnel, and a filter flask from the stockroom. Weigh the copper metal on the top-loading or triple-beam balance to 0.1 g.

Put the metal in a 150-mL beaker and **under a hood** add 5 mL 15 M HNO_3. **CAUTION:** *Caustic reagent.*

Brown NO_2 gas will be evolved and an acidic blue solution of $Cu(NO_3)_2$ produced. If it is necessary, you may warm the beaker gently with a Bunsen burner to dissolve all of the copper. When all of the copper is in solution, add 50 mL water to the solution and allow it to cool.

Weigh out about 5 grams of sodium carbonate in a small beaker on a rough balance. Add small amounts of the Na_2CO_3 to the solution with your spatula, adding the solid as necessary when the evolution of CO_2 subsides. Stir the solution to expose it to the solid. When the acid is neutralized, a blue-green precipitate of $CuCO_3$ will begin to form. At that point, add the rest of the Na_2CO_3, stirring the mixture well to ensure complete precipitation of the copper carbonate.

Transfer the precipitate to the Buchner funnel and use suction to remove the excess liquid. Use your rubber policeman and a spray from your wash bottle to make a complete transfer of the solid. Wash the precipitate well with distilled water with suction on, then let it remain on the filter paper with suction on for a minute or two.

Remove the filter paper from the funnel and transfer the solid $CuCO_3$ to the 150-mL beaker. Add 10 mL water then 30 mL 6 M HCl slowly to the solid, stirring continuously. When the $CuCO_3$ has all dissolved, add 1.5 g Cu turnings to the beaker and cover it with a watch glass.

Heat the mixture in the beaker to the boiling point and keep it at that temperature, just simmering, for about 10 minutes. It may be that the dark-colored solution that forms will clear to a yellow color before that time is up, and if it does, you may stop heating and proceed to the next step.

While the mixture is heating, put 150 mL distilled water in a 400-mL beaker and put the beaker in an ice bath. Cover the beaker with a watch glass. After you have heated the acidic $Cu-CuCl_2$ mixture for 10 minutes or as soon as it turns light colored, carefully decant the hot liquid into the beaker of water, taking care not to transfer any of the excess Cu metal to the beaker. White crystals of CuCl should form. Continue to cool the beaker in the ice bath to promote crystallization and to increase the yield of solid.

Cool 25 mL distilled water, to which you have added five drops 6 M HCl, in an ice bath. Put 20 mL acetone into a small beaker. Filter the crystals of CuCl in the Buchner funnel using suction. Swirl the beaker to aid in transferring the solid to the funnel. Just as the last of the liquid is being pulled through, wash the CuCl with 1/3 of the acidified cold water. Rinse the last of the CuCl into the funnel with another portion of the water and use the final 1/3 to rewash the solid. Turn off suction and add 1/2 of the acetone to the funnel; wait about 10 seconds and turn on the suction. Repeat this operation with the other half of the acetone. Draw air through the sample for a few minutes to dry it. If you have properly washed the solid, it will be pure white; if the moist compound is allowed to come into contact with air, it will tend to turn pale green, due to oxidation of Cu(I) to Cu(II). Weigh the CuCl in a previously weighed beaker to 0.1 g. Show your sample to your instructor for evaluation.

DISPOSAL OF REAGENTS. The contents of the suction flask after the first filtration (of $CuCO_3$) can be discarded down the sink. The contents after the second filtration include an appreciable amount of copper, so they should be put in the waste crock.

Determination of the Formula of Copper(I) Chloride (Optional)

There are many ways to prove that the formula of the copper compound you have prepared is indeed CuCl. The following procedure is easy to carry out and gives good results.

Dry the copper compound by putting it under a heat lamp or in a drying oven at 110°C for ten minutes. Weigh out about 0.1 g of the sample accurately on the analytical balance, using a weighed 50-mL beaker as a container. Dissolve the sample in 5 mL of 6 M HNO_3. Add 10 mL of distilled water, mix, and transfer the solution to a clean 100-mL volumetric flask. Use several 5-mL portions of distilled water to rinse the remaining solution from the beaker into the flask. Fill the flask to the mark with distilled water. Stopper the flask and invert it about 20 times to ensure that the solution in the flask is thoroughly mixed.

Use a clean 10-mL pipet to transfer a 10-mL aliquot of the solution to a clean, dry 50-mL beaker. Add 10 mL 6 M NH_3 to the beaker, using the 10-mL pipet, after you have rinsed it with 6 M NH_3. This will convert any Cu^{2+} in the solution to deep blue $Cu(NH_3)_4^{2+}$. After mixing, measure the absorbance of the solution of $Cu(NH_3)_4^{2+}$ at a wavelength of 575 nm. Determine $[Cu(NH_3)_4^{2+}]$ from a graph that is furnished to you or by comparison with a standard solution. Prepare the standard by accurately weighing about 0.06 g of copper turnings and putting them into a 50-mL beaker. Dissolve the copper in 5 mL of 6 M HNO_3, and proceed as you did with the solution of the copper compound in the previous paragraph. Measure the absorbance of the standard at 575 nm. Calculate $[Cu(NH_3)_4^{2+}]$ in the solution of the compound by assuming that Beer's Law holds. Knowing that concentration, and the fact that the sample was diluted to a volume of 200 mL, calculate the number of moles of Cu in the sample. Then calculate the number of grams of Cu in the sample, and the number of grams of Cl by difference. Use that value to find the number of moles of Cl, and from the mole ratio Cu:Cl obtain the formula of the compound.

Data and Results: Preparation of CuCl

Mass of Cu sample   _____ g

Mass of beaker   _____ g

Mass of beaker plus CuCl   _____ g

Mass of CuCl prepared   _____ g

Theoretical yield   _____ g

Percentage yield   _____ %

Determination of the Formula of Copper(I) Chloride (Optional)

Mass of Cu(I) chloride sample   _____ g

Absorbance of solution of $Cu(NH_3)_4^{2+}$ _____

Mass of Cu turnings   _____ g

Absorbance of standard solution   _____

No. moles Cu in turnings   _____ moles

All of the copper in the turnings is converted to $Cu(NH_3)_4^{2+}$
in a solution whose total volume would be 200 mL.

$[Cu(NH_3)_4^{2+}]$ in standard solution   _____ M

If Beer's Law holds,

$$[Cu(NH_3)_4^{2+}] \text{ in solution of sample} = [Cu(NH_3)_4^{2+}] \text{ in standard} \times \frac{\text{Abs of sample}}{\text{Abs of standard}}$$

$[Cu(NH_3)_4^{2+}]$ in solution of sample   _____ M

No. moles Cu in sample (volume = 0.200 L)   _____ moles

No. grams Cu in sample   _____ g

(continued on following page)

(continued)

No. grams Cl in sample (sample mass = mass Cu + mass Cl) _____ g

No. moles Cl in sample _____ g

Mole ratio Cl : Cu _____

Formula of prepared compound (rounded to nearest integers) _____

Advanced Study Assignment: Preparation of Copper(I) Chloride

1. The Cu^{2+} ions in this experiment are produced from the reaction of 1.0 g of copper turnings with excess nitric acid. How many moles of Cu^{2+} are produced?

    _____ moles Cu^{2+}

2. Why isn't hydrochloric acid used in a direct reaction with copper to prepare the $CuCl_2$ solution?

3. How many grams of metallic copper are required to react with the number of moles of Cu^{2+} calculated in Problem 1 to form the CuCl? The overall reaction can be taken to be:
    $$Cu^{2+}(aq) + 2\ Cl^-(aq) + Cu(s) \rightarrow 2\ CuCl(s)$$

    _____ g Cu

4. What is the maximum mass of CuCl that can be prepared from the reaction sequence of this experiment using 1.0 g of Cu turnings to prepare the Cu^{2+} solution?

    _____ g CuCl

5. Indicate all the species, in proper sequence, to which the copper in the turnings is transformed during the course of the synthesis of CuCl.

DEVELOPMENT OF A SCHEME FOR QUALITATIVE ANALYSIS

In many of the previous experiments in this book you were asked to find out how much of a given species is present in a sample. You may have determined the percent chloride in an unknown solid, the molarity of an NaOH solution, and the amount of calcium ion in a sample of hard water. All of these experiments fall into that part of chemistry called quantitative analysis.

Sometimes chemists are interested more in the nature of the species in a sample than the amount of those species. In that sort of problem we find out what the sample contains but not how much. For example, in Experiment 11 students are asked to determine which alkaline earth halide is present in a solution. That experiment is one involving qualitative analysis. This and the next three experiments deal with the qualitative analysis of solutions containing various metallic cations.

The procedures in qualitative analyses of this sort involve using precipitation reactions to remove the cations sequentially from a mixture. If the precipitate can contain only one cation under the conditions that prevail, that precipitate serves to prove the presence of that cation. If the precipitate may contain several cations, it can be dissolved and further resolved in a series of steps that may include acid-base, complex ion formation, redox, or other precipitation reactions. The ultimate result is a resolution of the sample into fractions that can contain only one cation, whose presence is established by formation of a characteristic precipitate or a colored complex ion.

In this experiment you will be asked to develop a scheme for the qualitative analysis of four cations, using this systematic approach. The behavior of the cations toward a set of common test reagents differs from one cation to another and furnishes the basis for their separation.

The cations we will study are Ba^{2+}, Mg^{2+}, Cd^{2+}, and Al^{3+}. The test reagents we will use are 1 M Na_2SO_4, 1 M Na_2CO_3, 6 M NaOH, 6 M NH_3, and 6 M HNO_3. These reagents furnish anions or molecules that will precipitate or form complex ions with the cations. So you may observe formation of insoluble sulfates, carbonates, and hydroxides (the last with either NaOH or NH_3). In addition you may form complexes with OH^- and NH_3 when 6 M NaOH or 6 M NH_3 is added to the cation-containing solutions. The complexes may be quite stable, stable enough to prevent precipitation of an otherwise insoluble salt on addition of a particular anion. The complexes, however, are all unstable in excess acid and can be broken down on addition of 6 M HNO_3, releasing the cation for further reactions.

In the first part of the experiment you will observe the behavior of the four cations in the presence of one or more of the reagents we have listed. On the basis of your observations you can set up the scheme for identifying the cations in a mixture. After testing the scheme with a known containing all of the cations, you will be given an unknown to analyze.

Experimental Procedure

To four small test tubes add 1 mL of 0.1 M solutions of the nitrates or chlorides of Ba^{2+}, Mg^{2+}, Cd^{2+}, and Al^{3+}, one solution to a test tube (depth of 1 cm $\cong$ 1 mL). To each solution add one drop of 1 M Na_2SO_4, and stir with your stirring rod. Keep the rod in a 400-mL beaker of distilled water between tests. If a precipitate forms, put the formula of the precipitate in the box on the Data page corresponding to the cation-SO_4^{2-} pair, to indicate that the sulfate of that cation is insoluble in water. Then add 1 mL more of the Na_2SO_4 to each test tube. If the precipitate dissolves on stirring, the cation forms a complex ion with sulfate ion. Under such conditions, put the formula of the complex ion in the box. In this experiment you can assume that any cations that form complex ions have a coordination number of four. To any precipitates or complex ions, add 6 M HNO_3, drop by drop, until the solution is acidic to litmus (blue to red). If the precipitate dissolves, note that with an *A*, meaning that the precipitate dissolves in acid. In the case of complexes, the precipitate that originally formed may reprecipitate if the ligands react with acid, and then dissolve again when the solution becomes acidic. If it reprecipitates, indicate that with a *P*, and as before use an *A* if the precipitate dissolves when the solution becomes acidic.

Pour the contents of the four tubes into a beaker and rinse the tubes with distilled water. Repeat the tests, first with 1 M Na_2CO_3, then with 6 M NaOH, and finally with 6 M NH_3. Although in most cases you will not observe formation of complex ions, you will see a few, and it is important not to miss them. One drop of reagent typically will produce a precipitate if the cation-anion compound is insoluble in water. The complex may be very stable and form readily in excess reagent. You may not get a precipitate reforming when you add HNO_3. If the solution becomes acidic, and you saw no precipitate or a faint one, slowly add 1 M Na_2CO_3. If the carbonate comes down, write its formula in the box.

When your table is complete you should be able to use it to state whether the sulfate, carbonate, and hydroxide of each of the cations is insoluble in water, and whether it dissolves in acid. You should also know whether the cation forms a complex ion with SO_4^{2-}, CO_3^{2-}, OH^-, or NH_3, and whether the complex is destroyed by acid.

Now, given the solubility and reaction data you have obtained, your problem is to devise a step-by-step procedure for establishing whether each cation is present in a mixture. If you think about it for a while, several possible approaches should occur to you. As you construct your scheme, there are a few things to keep in mind besides the solubility and reaction data.

1. To separate a precipitate from a solution, you can use a centrifuge. Decant the solution into a test tube for use in further steps. See Appendix IV for some suggestions regarding procedures in qualitative analysis.
2. If a precipitate can contain only one cation, its presence serves to prove the presence of that cation. If it may involve more than one cation, it must be further resolved. In that case, the precipitate must be washed free of any cations that did not precipitate in that step because those cations would possibly interfere with later steps. To clean a precipitate, wash it twice with a 1:1 mixture of water and the precipitating reagent, stirring before centrifuging out the wash liquid.
3. pH is important. Your original tests were made with the cations in a water solution. If you want to obtain a precipitate, the solution must have a pH where that precipitate can form. 6 M HNO_3 or 6 M NaOH can be used to bring a mixture to a pH of about 7.

    When you have your separation scheme in mind, describe it by constructing a flow diagram. The design of a flow diagram is discussed in the Advance Study Assignment. On the flow diagram the formulas of all species involving the cations should be given at the beginning and end of each step. Reagents are shown alongside the line connecting reactants and products.

    When you have completed your flow diagram, test your scheme with a known containing all four cations. If your scheme works, show your flow diagram to your instructor, who will give you an unknown to analyze.

On completing the experiment, pour the contents of the beaker used for reaction products into the waste crock unless directed otherwise by your instructor.

Data: Development of a Scheme for Qualitative Analysis

Table of Solubility Properties

	Ba^{2+}	Mg^{2+}	Cd^{2+}	Al^{3+}
1 M Na_2SO_4				
1 M Na_2CO_3				
6 M NaOH				
6 M NH_3				

Key: *No entry = soluble on mixing reagents*
Top formula = precipitate insoluble in water
Second formula = complex ion that forms in excess reagent
Bottom formula = carbonate that precipitates from acidified complex
A = precipitate dissolves in acid
P = precipitate reforms when complex is slowly acidified

Flow Diagram for Separation Scheme:

Observations on known:

Observations on unknown:

Cations present in unknown: _____ _____ _____ _____

Unknown no. _____

Advance Study Assignment: Development of a Scheme for Qualitative Analysis

Qualitative analysis schemes can be summarized by flow diagrams. The flow diagram for a scheme that might be used to analyze a mixture that could contain Cu^{2+}, Pb^{2+}, and Sn^{2+} is shown below:

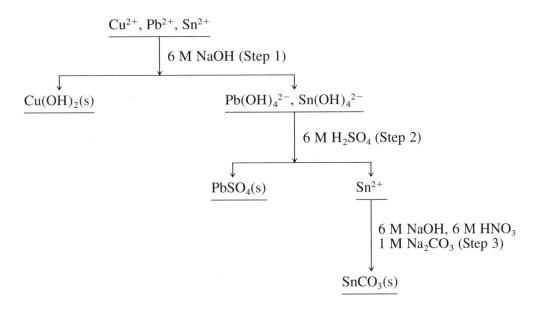

The meaning of the flow diagram is almost obvious. In the first step, 6 M NaOH is added in excess. $Cu(OH)_2$ precipitates, and Pb^{2+} and Sn^{2+} remain in solution because of the formation of hydroxo-complex ions. $Cu(OH)_2$ is removed by centrifuging. In Step 2 the solution is made acidic with H_2SO_4, and $PbSO_4$ precipitates after the complex is destroyed by the H^+ ion in the acid. $PbSO_4$ is removed by centrifuging. In Step 3 pH control is used to bring the pH to about 7. Then addition of Na_2CO_3 precipitates $SnCO_3$.

1. Construct the flow diagram for the separation scheme for a solution containing Ag^+, Ni^{2+}, and Zn^{2+}. The steps in the procedure are as follows:

    Step 1. Add 6 M HCl to precipitate Ag^+ as AgCl. Ni^{2+} and Zn^{2+} are not affected. Centrifuge out the AgCl.

    Step 2. Add 6 M NaOH in excess, precipitating $Ni(OH)_2(s)$ and converting Zn^{2+} to the $Zn(OH)_4^{2-}$ complex ion. Centrifuge out the $Ni(OH)_2$.

    Step 3. Neutralize the solution with 6 M HCl to pH = 7. Add 1 M Na_2CO_3, precipitating $ZnCO_3$.

Experiment 33

QUALITATIVE ANALYSIS OF GROUP I CATIONS

PRECIPITATION AND SEPARATION OF GROUP I IONS

The chlorides of Pb^{2+}, Hg_2^{2+}, and Ag^+ are all insoluble in cold water. They can be removed as a group from solution by the addition of HCl. The reactions that occur are simple precipitations and can be represented by the equations:

$$Ag^+(aq) + Cl^-(aq) \rightarrow AgCl(s) \tag{1}$$

$$Pb^{2+}(aq) + 2\,Cl^-(aq) \rightarrow PbCl_2(s) \tag{2}$$

$$Hg_2^{2+}(aq) + 2\,Cl^-(aq) \rightarrow Hg_2Cl_2(s) \tag{3}$$

It is important to add enough HCl to ensure complete precipitation, but not too large an excess. In concentrated HCl solution these chlorides tend to dissolve, producing chloro-complexes such as $AgCl_2^-$.

Lead chloride is separated from the other two chlorides by heating with water. The $PbCl_2$ dissolves in hot water by the reverse of Reaction 2:

$$PbCl_2(s) \rightarrow Pb^{2+}(aq) + 2\,Cl^-(aq) \tag{4}$$

Once Pb^{2+} has been put into solution, we can check for its presence by adding a solution of K_2CrO_4. The chromate ion, CrO_4^{2-}, gives a yellow precipitate with Pb^{2+}:

$$Pb^{2+}(aq) + CrO_4^{2-}(aq) \rightarrow PbCrO_4(s) \tag{5}$$
$$\text{yellow}$$

The other two insoluble chlorides, $AgCl$ and Hg_2Cl_2, can be separated by adding aqueous ammonia. Silver chloride dissolves, forming the complex ion $Ag(NH_3)_2^+$:

$$AgCl(s) + 2\,NH_3(aq) \rightarrow Ag(NH_3)_2^+(aq) + Cl^-(aq) \tag{6}$$

Ammonia also reacts with Hg_2Cl_2 via a rather unusual oxidation-reduction reaction. The products include finely divided metallic mercury, which is black, and a compound of formula $HgNH_2Cl$, which is white:

$$Hg_2Cl_2(s) + 2\,NH_3(aq) \rightarrow Hg(l) + HgNH_2Cl(s) + NH_4^+(aq) + Cl^-(aq) \tag{7}$$
$$\quad\text{white} \qquad\qquad\qquad \text{black} \qquad \text{white}$$

As this reaction occurs, the solid appears to change color, from white to black or gray.

The solution containing $Ag(NH_3)_2^+$ needs to be further tested to establish the presence of silver. The addition of a strong acid (HNO_3) to the solution destroys the complex ion and reprecipitates silver chloride. We may consider that this reaction occurs in two steps:

$$Ag(NH_3)_2^+(aq) + 2 H^+(aq) \rightarrow Ag^+(aq) + 2 NH_4^+(aq)$$
$$\underline{Ag^+(aq) + Cl^-(aq) \rightarrow AgCl(s)}$$
$$Ag(NH_3)_2^+(aq) + 2 H^+(aq) + Cl^-(aq) \rightarrow AgCl(s) + 2 NH_4^+(aq)$$
$$\text{white}$$

(8)

Experimental Procedure

See Appendix IV for some suggestions regarding procedures in qualitative analysis.

Step 1. Precipitation of the Group I Cations. To gain familiarity with the procedures used in qualitative analysis we will first analyze a known Group I solution, made by mixing equal volumes of 0.1 M $AgNO_3$, 0.2 M $Pb(NO_3)_2$, and 0.1 M $Hg_2(NO_3)_2$.

Add two drops of 6 M HCl to 1 mL of the known solution in a small test tube. (1 mL $\cong$ 1 cm depth in the tube.) Mix with your stirring rod. Centrifuge the mixture, making sure there is a blank test tube containing about the same amount of water in the opposite opening in the centrifuge. Add one more drop of the 6 M HCl to test for completeness of precipitation. Centrifuge again if necessary. Decant the supernatant liquid into another test tube and save it for further tests if cations from other groups may be present. The precipitate will be white and will contain the chlorides of the Group I cations.

Step 2. Separation of Pb^{2+}. Wash the precipitate with 1 or 2 mL of water. Stir, centrifuge, and decant the liquid, which may be discarded.

Add 2 mL distilled water from your wash bottle to the precipitate in the test tube, and place the test tube in a 250-mL beaker that is about half full of boiling water. Leave the test tube in the bath for a minute or two, stirring occasionally with a glass rod. This will dissolve most of the $PbCl_2$, but not the other two chlorides. Centrifuge the hot mixture, and decant the hot liquid into a test tube. Save the remaining precipitate for further tests.

Step 3. Identification of Pb^{2+}. Add one drop of 6 M acetic acid and a drop or two of 1 M K_2CrO_4 to the solution from Step 2. If Pb^{2+} is present, a bright yellow precipitate of $PbCrO_4$ will form.

Step 4. Separation and Identification of Hg_2^{2+}. To the precipitate from Step 2 add 1 mL 6 M NH_3 and stir thoroughly. Centrifuge the mixture and decant the liquid into a test tube. A gray or black precipitate, produced by reaction of Hg_2Cl_2 with ammonia, proves the presence of Hg_2^{2+}.

Step 5. Identification of Ag^+. Add 6 M HNO_3 to the solution from Step 4 until it is acidic to litmus paper. It will take about 1 mL. Test for acidity by dipping the end of your stirring rod in the solution then touching it to a piece of blue litmus paper (red in acidic solution). If Ag^+ is present, in the acidified solution it will precipitate as white AgCl.

Step 6. When you have completed the tests on the known solution, obtain an unknown and analyze it for the possible presence of Ag^+, Pb^{2+}, and Hg_2^{2+}.

All reaction products in this experiment should be dealt with as directed by your instructor.

Flow Diagrams

It is possible to summarize the directions for analysis of the Group I cations in a flow diagram. In the diagram, successive steps in the procedure are linked with arrows. Reactant cations or reactant substances containing the ions are at one end of each arrow and products formed are at the other end. Reagents and conditions used to carry out each step are placed alongside the arrows. A partially completed flow diagram for the Group I ions follows:

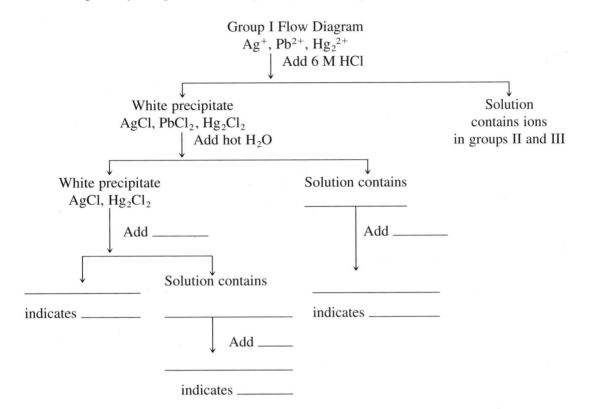

Group I Flow Diagram
Ag^+, Pb^{2+}, Hg_2^{2+}
Add 6 M HCl

White precipitate
$AgCl$, $PbCl_2$, Hg_2Cl_2
Add hot H_2O

Solution
contains ions
in groups II and III

White precipitate
$AgCl$, Hg_2Cl_2
Add _____

Solution contains
Add _____

Solution contains
Add _____

indicates _____

indicates _____

indicates _____

You will find it useful to construct flow diagrams for each of the cation groups. You can use such diagrams in the laboratory as brief guides to procedure, and you can use them to record your observations on your known and unknown solutions.

Observations and Report Sheet: Qualitative Analysis of Group I Cations

Flow Diagram for Group I

Observations on known (record on diagram if different from those on prepared diagram).

Observations on unknown (record on diagram in colored pencil to distinguish from observations on known).

Unknown no. _____

Ions reported present _____ _____ _____

Experiment 33 *Name* ______________________ *Section* __________

Advance Study Assignment: Group I Cations

1. On the report sheet, complete the flow diagram for the separation and identification of the ions in Group I.

2. Write balanced net ionic equations for the following reactions:

a. The precipitation of the chloride of Hg_2^{2+} in Step 1.

b. The formation of a yellow precipitate in Step 3.

c. The formation of a black precipitate in Step 4.

d. The reaction that occurs in Step 5.

3. A solution may contain Ag^+, Pb^{2+}, and Hg_2^{2+}. A white precipitate forms on addition of 6 M HCl. The precipitate is partially soluble in hot water; the residue dissolves on addition of 6 M NH_3. Which of the ions are present, which are absent, and which remain undetermined? State your reasoning. (In ''paper'' unknowns such as this one, confirmatory tests are often omitted. The partial solubility of the precipitate in hot water implies that one of the ions is present, and so we report it present.)

Present ______________

Absent ______________

In doubt ______________

Experiment 34

QUALITATIVE ANALYSIS OF GROUP II CATIONS

PRECIPITATION AND SEPARATION OF GROUP II IONS

The sulfides of the four Group II ions, Bi^{3+}, Sn^{4+}, Sb^{3+}, and Cu^{2+}, are insoluble at a pH of 0.5. The solution is adjusted to this pH and then saturated with H_2S, which precipitates Bi_2S_3, SnS_2, Sb_2S_3, and CuS. The reaction with Bi^{3+} is typical:

$$2\,Bi^{3+}(aq) + 3\,H_2S(aq) \rightarrow Bi_2S_3(s) + 6\,H^+(aq) \tag{1}$$
$$\text{black}$$

Saturation with H_2S could be achieved by simply bubbling the gas from a generator through the solution. A more convenient method, however, is to heat the acid solution after adding a small amount of thioacetamide. This compound, CH_3CSNH_2, hydrolyzes when heated in water solution to liberate H_2S:

$$CH_3CSNH_2(aq) + 2\,H_2O \rightarrow H_2S(aq) + CH_3COO^-(aq) + NH_4^+(aq) \tag{2}$$

Using thioacetamide as the precipitating reagent has the advantage of minimizing odor problems and giving denser precipitates.

The four insoluble sulfides can be separated into two subgroups by extracting with a solution of sodium hydroxide. The sulfides of tin and antimony dissolve, forming hydroxo-complexes:

$$SnS_2(s) + 6\,OH^-(aq) \rightarrow Sn(OH)_6^{2-}(aq) + 2\,S^{2-}(aq) \tag{3}$$

$$Sb_2S_3(s) + 8\,OH^-(aq) \rightarrow 2\,Sb(OH)_4^-(aq) + 3\,S^{2-}(aq) \tag{4}$$

Since Cu^{2+} and Bi^{3+} do not readily form hydroxo-complexes, CuS and Bi_2S_3 do not dissolve in solutions of $NaOH$.

The solution containing the $Sb(OH)_4^-$ and $Sn(OH)_6^{2-}$ complex ions is treated with HCl and thioacetamide. The H^+ ions of the strong acid HCl destroy the hydroxocomplexes; the free cations then reprecipitate as the sulfides. The reaction with $Sn(OH)_6^{2-}$ may be written as:

$$Sn(OH)_6^{2-}(aq) + 6\,H^+(aq) \rightarrow Sn^{4+}(aq) + 6\,H_2O$$

$$\underline{Sn^{4+}(aq) + 2\,H_2S(aq) \rightarrow SnS_2(s) + 4\,H^+(aq)}$$
$$Sn(OH)_6^{2-}(aq) + 2\,H^+(aq) + 2\,H_2S(aq) \rightarrow SnS_2(s) + 6\,H_2O \tag{5}$$
$$\text{tan}$$

The $Sb(OH)_4^-$ ion behaves in a very similar manner, being converted first to Sb^{3+} and then to Sb_2S_3. The Sb_2S_3 and SnS_2 are then dissolved as chloro-complexes in hydrochloric acid and their presence is confirmed by appropriate tests.

The confirmatory test for tin takes advantage of the two oxidation states, $+2$ and $+4$, of the metal. Aluminum is added to reduce Sn^{4+} to Sn^{2+}:

$$2\,Al(s) + 3\,Sn^{4+}(aq) \rightarrow 2\,Al^{3+}(aq) + 3\,Sn^{2+}(aq) \tag{6}$$

The presence of Sn^{2+} in this solution is detected by adding mercury(II) chloride, $HgCl_2$. This brings about another oxidation-reduction reaction:

$$Sn^{2+}(aq) + 2\ Hg^{2+}(aq) + 2\ Cl^-(aq) \rightarrow Sn^{4+}(aq) + Hg_2Cl_2(s) \qquad (7)$$
$$\text{white}$$

Formation of a white precipitate of insoluble Hg_2Cl_2 confirms the presence of tin. (The precipitate may be grayish because of the formation of finely divided Hg by an oxidation-reduction reaction similar to Reaction 7.)

The Sb^{3+} ion is difficult to confirm in the presence of Sn^{4+}; the colors of the sulfides of these two ions are similar. To prevent interference by Sn^{4+}, the solution to be tested for Sb^{3+} is first treated with oxalic acid. This forms a very stable oxalato-complex with Sn^{4+}, $Sn(C_2O_4)_3^{2-}$. Treatment with H_2S then gives a bright-orange precipitate of Sb_2S_3 if antimony is present:

$$2Sb^{3+}(aq) + 3\ H_2S(aq) \rightarrow Sb_2S_3(s) + 6\ H^+(aq) \qquad (8)$$
$$\text{orange}$$

As pointed out earlier, CuS and Bi_2S_3 are insoluble in NaOH solutions, and they do not dissolve in hydrochloric acid. However, these two sulfides can be brought into solution by treatment with the oxidizing acid, HNO_3. The reactions that occur are of the oxidation-reduction type. The NO_3^- ion is reduced, usually to NO_2; S^{2-} ions are oxidized to elementary sulfur and the cation, Cu^{2+} or Bi^{3+}, is brought into solution. The equations are:

$$CuS(s) + 4\ H^+(aq) + 2\ NO_3^-(aq) \rightarrow Cu^{2+}(aq) + S(s) + 2\ NO_2(g) + 2H_2O \qquad (9)$$

$$Bi_2S_3(s) + 12\ H^+(aq) + 6\ NO_3^-(aq) \rightarrow 2\ Bi^{3+}(aq) + 3S(s) + 6\ NO_2(g) + 6\ H_2O \qquad (10)$$

The two ions, Cu^{2+} and Bi^{3+}, are easily separated by the addition of aqueous ammonia. The Cu^{2+} ion is converted to the deep-blue complex, $Cu(NH_3)_4^{2+}$:

$$Cu^{2+}(aq) + 4\ NH_3(aq) \rightarrow Cu(NH_3)_4^{2+}(aq) \qquad (11)$$
$$\text{deep blue}$$

The reaction of ammonia with Bi^{3+} is quite different. The OH^- ions produced by the reaction of NH_3 with water precipitate Bi^{3+} as $Bi(OH)_3$. We can consider that the reaction occurs in two steps:

$$3\ NH_3(aq) + 3\ H_2O \rightleftarrows 3\ NH_4^+(aq) + 3\ OH^-(aq)$$

$$\frac{Bi^{3+}(aq) + 3\ OH^-(aq) \rightarrow Bi(OH)_3(s)}{Bi^{3+}(aq) + 3\ NH_3(aq) + 3\ H_2O \rightarrow Bi(OH)_3(s) + 3\ NH_4^+(aq)} \qquad (12)$$
$$\text{white}$$

To confirm the presence of Bi^{3+}, the precipitate of $Bi(OH)_3$ is dissolved by treating with hydrochloric acid:

$$Bi(OH)_3(s) + 3\ H^+(aq) \rightarrow Bi^{3+}(aq) + 3\ H_2O \qquad (13)$$

The solution formed is poured into distilled water. If Bi^{3+} is present, a white precipitate of bismuth oxychloride, BiOCl, will form:

$$Bi^{3+}(aq) + H_2O + Cl^-(aq) \rightarrow BiOCl(s) + 2\ H^+(aq) \qquad (14)$$
$$\text{white}$$

Experimental Procedure

Step 1. Adjustment of pH Prior to Precipitation. Pour a 1-mL sample of the "known" solution for Group II, containing equal volumes of 0.1 M solutions of the nitrates or chlorides of Sn^{4+}, Sb^{3+}, Cu^{2+}, and Bi^{3+}, into a small test tube. Prepare a boiling water bath, using a 250-mL beaker about 2/3 full of water. Use another beaker full of water as the storage and rinsing place for your stirring rods. Prepare a pH test paper by making ten spots of methyl violet indicator on a piece of filter paper, about one drop to a spot. Let the paper dry in the air.

The Group II known will be very acidic because of the presence of HCl, which is necessary to keep the salts of Bi^{3+}, Sn^{4+}, and Sn^{3+} in solution. Add 6 M NH_3, drop by drop, until the solution, after stirring, produces a violet spot on the pH test paper; test for pH by dipping your stirring rod in the solution and touching it to the paper. (A precipitate will probably form during this step because of the formation of insoluble salts of the Group II cations.) Then add 1 drop of 6 M HCl for each milliliter of solution. This should bring the pH of the solution to about 0.5. Test the pH, using the test paper. At a pH of 0.5, methyl violet will have a blue-green color. Compare your spot with that made by putting a drop of 0.3 M HCl on the test paper (pH = 0.5). Adjust the pH of your solution as necessary by adding HCl or NH_3, until the pH test gives about the right color on the paper. If you have trouble deciding on the color, centrifuging out the precipitate may help. When you have established the proper pH, add 1 mL 1 M thioacetamide to the solution and stir.

Step 2. Precipitation of the Group II Sulfides. Heat the test tube in the boiling water bath for at least five minutes. **CAUTION:** *Small amounts of H_2S will be liberated; this gas is toxic, so avoid inhaling it unnecessarily.*

In the presence of Group II ions, a precipitate will form: typically, its color will be initially light, gradually darkening, and finally becoming black. Continue to heat the tube for at least two minutes after the color has stopped changing. Cool the test tube under the water tap and let it stand for a minute or so. Centrifuge out the precipitate and decant the solution, which will contain any Group III ions if they are present, into a test tube. Test the solution for completeness of precipitation by adding two drops of thioacetamide and letting it stand for a minute. If a precipitate forms, add a few drops of thioacetamide and heat again in the water bath. Combine the two batches of precipitate. Wash the precipitate with 2 mL 1 M NH_4Cl solution; stir thoroughly. Centrifuge and discard the wash into a waste beaker.

Step 3. Separation of the Group II Sulfides into Two Subgroups. To the precipitate from Step 2, add 2 mL 1 M NaOH. Heat in the water bath, with stirring, for two minutes. Any SnS_2 or Sb_2S_3 should dissolve. The residue will typically be dark and may contain CuS and Bi_2S_3. Centrifuge and decant the yellow liquid into a test tube. Wash the precipitate twice with 2 mL water and a few drops of 1 M NaOH. Stir, centrifuge, and decant, discarding the liquid each time into a waste beaker. To the precipitate add 2 mL 6 M HNO_3 and put the test tube aside (Step 8). At this point the tin and antimony are present in the solution as complex ions, and the copper and bismuth are in the sulfide precipitate.

Step 4. Reprecipitation of SnS_2 and Sb_2S_3. To the yellow liquid from Step 3 add 6 M HCl drop by drop until the mixture is just acidic to litmus. Upon acidification, the tin and antimony will again precipitate as orange sulfides. Add five drops of 1 M thioacetamide and heat in the water bath for two minutes to complete the precipitation. Centrifuge, and decant the liquid, which may be discarded.

Step 5. Dissolving SnS$_2$ and Sb$_2$S$_3$ in Acid Solution. Add 2 mL 6 M HCl to the precipitate from Step 4, and heat in the water bath for a minute or two to dissolve the precipitate. Transfer the solution to a 30-mL beaker, and boil it, gently, for about a minute, to drive out the H$_2$S. Add 1 mL 6 M HCl and 1 mL water and pour the liquid back into a small test tube. If there is an insoluble residue, centrifuge it out and decant the liquid into a test tube.

Step 6. Confirmation of the Presence of Tin. Pour half of the solution from Step 5 into a test tube and add 2 mL 6 M HCl and a 1-cm length of 24-gauge aluminum wire. Heat the test tube in the water bath to promote reaction of the Al and production of H$_2$. In this reducing medium, any tin present will be converted to Sn^{2+} and any antimony to the metal, which will appear as black specks. Heat the tube for two minutes after *all* of the wire has reacted. Centrifuge out any solid and decant the liquid into a test tube. To the liquid add a drop or two of 0.1 M HgCl$_2$. A white or gray cloudiness, produced as Hg$_2$Cl$_2$ or Hg slowly forms, establishes the presence of tin.

Step 7. Confirmation of the Presence of Antimony. To the other half of the solution from Step 5, add 1 M NaOH to bring the pH to 0.5. Add 2 mL water and about 0.5 g of oxalic acid, and stir until no more crystals dissolve. Oxalic acid forms a very stable complex with the Sn^{4+} ion. Add 1 mL 1 M thioacetamide and put the test tube in the water bath. The formation of a red-orange precipitate of Sb$_2$S$_3$ confirms the presence of antimony.

Step 8. Dissolving the CuS and Bi$_2$S$_3$. Heat the test tube containing the precipitate from Step 3 in the hot water bath. Any sulfides that have not already dissolved should go into solution in a minute or two, possibly leaving some insoluble sulfur residue. Continue heating until no further reaction appears to occur, at least two minutes after the initial changes. Centrifuge and decant the solution, which may contain Cu^{2+} and Bi^{3+}, into a test tube. Discard the residue.

Step 9. Confirmation of the Presence of Copper. To the solution from Step 8 add 6 M NH$_3$ dropwise until the mixture is just basic to litmus. Add 0.5 mL more. Centrifuge out any white precipitate that forms, and decant the liquid into a test tube. If the liquid is deep blue, the color is due to the Cu(NH$_3$)$_4^{2+}$ ion, and copper is present.

Step 10. Confirmation of the Presence of Bismuth. Wash the precipitate from Step 9, which probably contains bismuth, with 1 mL water and 0.5 mL 6 M NH$_3$. Stir, centrifuge, and discard the wash. To the precipitate add 0.5 mL 6 M HCl and 0.5 mL water. Stir to dissolve any Bi(OH)$_3$ that is present. Add the solution drop by drop, to 400 mL water in a 600-mL beaker. A white cloudiness, caused by slow precipitation of BiOCl, confirms the presence of bismuth.

Step 11. When you have completed the analysis of your known, obtain a Group II unknown and test it for the presence of Sn^{4+}, Sb^{3+}, Cu^{2+}, and Bi^{3+}.

DISPOSAL OF REAGENTS. As you complete each part of the experiment, put the waste products in a beaker. At the end of the experiment, pour the contents of the beaker into a waste crock, unless directed otherwise by your instructor.

Name _____   *Section* _____

Observations and Report Sheet: Qualitative Analysis of Group II Cations

Flow Diagram for Group II

Observations on known (record on diagram if different from those on prepared diagram).

Observations on unknown (record on diagram in colored pencil to distinguish from observations on known).

Unknown no. _____

Ions reported present _____   _____   _____   _____

Advance Study Assignment: Qualitative Analysis of Group II Cations

1. Prepare a complete flow diagram for the separation and identification of the Group II cations and put it on your report sheet.

2. Write balanced net ionic equations for the following reactions:

    a. Precipitation of the tin(IV) sulfide with H_2S.

    b. The confirmatory test for antimony.

    c. The confirmatory test for bismuth.

    d. The dissolving of Bi_2S_3 in hot nitric acid.

3. A solution that may contain Cu^{2+}, Bi^{3+}, Sn^{4+}, or Sb^{3+} ions is treated with thioacetamide in an acid medium. The black precipitate that forms is partly soluble in strongly alkaline solution. The precipitate that remains is soluble in 6 M HNO_3 and gives only a blue solution on treatment with excess NH_3. The alkaline solution, when acidified, produces an orange precipitate. On the basis of this information, which ions are present, which are absent, and which are still in doubt? (As with group I, evidence other than confirmatory tests may show the presence or absence of an ion.)

    Present _____

    Absent _____

    In doubt _____

Experiment 35

QUALITATIVE ANALYSIS OF GROUP III CATIONS

PRECIPITATION AND SEPARATION OF THE GROUP III IONS

Four ions in Group III are Cr^{3+}, Al^{3+}, Fe^{3+}, and Ni^{2+}. The first step in the analysis involves treating the solution with sodium hydroxide, NaOH, and sodium hypochlorite, NaOCl. The OCl^- ion oxidizes Cr^{3+} to CrO_4^{2-}:

$$2\,Cr^{3+}(aq) + 3\,OCl^-(aq) + 10\,OH^-(aq) \rightarrow 2\,CrO_4^{2-}(aq) + 3\,Cl^-(aq) + 5\,H_2O \quad (1)$$
$$\text{yellow}$$

The chromate ion, CrO_4^{2-}, stays in solution. The same is true of the hydroxo-complex ion $Al(OH)_4^-$, formed by the reaction of Al^{3+} with excess OH^-:

$$Al^{3+}(aq) + 4\,OH^-(aq) \rightarrow Al(OH)_4^-(aq) \quad (2)$$

In contrast, the other two ions in the group form insoluble hydroxides under these conditions:

$$Ni^{2+}(aq) + 2\,OH^-(aq) \rightarrow Ni(OH)_2(s) \quad (3)$$
$$\text{green}$$

$$Fe^{3+}(aq) + 3\,OH^-(aq) \rightarrow Fe(OH)_3(s) \quad (4)$$
$$\text{red}$$

The Ni^{2+} and Fe^{3+} ions, unlike Al^{3+}, do not readily form hydroxo-complexes. Unlike Cr^{3+}, they do not have a stable higher oxidation state and so are not oxidized by ClO^-.

To separate aluminum from chromium, the solution containing CrO_4^{2-} and $Al(OH)_4^-$ is first acidified. This destroys the hydroxo-complex of aluminum:

$$Al(OH)_4^-(aq) + 4\,H^+(aq) \rightarrow Al^{3+}(aq) + 4\,H_2O \quad (5)$$

Treatment with aqueous ammonia then gives a white gelatinous precipitate of aluminum hydroxide. We can think of this reaction as occurring in two steps:

$$3\,NH_3(aq) + 3\,H_2O \rightleftharpoons NH_4^+(aq) + 3\,OH^-(aq)$$

$$\underline{Al^{3+}(aq) + 3\,OH^-(aq) \rightarrow Al(OH)_3(s)}$$

$$Al^{3+}(aq) + 3\,NH_3(aq) + 3\,H_2O \rightarrow Al(OH)_3(s) + 3\,NH_4^+(aq) \quad (6)$$
$$\text{white}$$

The concentration of OH^- in dilute NH_3 is too low to form the $Al(OH)_4^-$ complex ion by Reaction 2.

The CrO_4^{2-} ion remains in solution after Al^{3+} has been precipitated. It can be tested for by precipitation as yellow $BaCrO_4$ by the addition of $BaCl_2$ solution:

$$Ba^{2+}(aq) + CrO_4^{2-}(aq) \rightarrow BaCrO_4(s) \quad (7)$$
$$\text{light yellow}$$

The precipitate of $BaCrO_4$ is dissolved in acid; the solution formed is then treated with hydrogen peroxide, H_2O_2. A deep-blue color is produced, because of the presence of a peroxo-compound, probably CrO_5. The reaction may be represented by the overall equation:

$$2\,BaCrO_4(s) + 4\,H^+(aq) + 4\,H_2O_2(aq) \rightarrow 2\,Ba^{2+}(aq) + 2\,CrO_5(aq) + 6\,H_2O \quad (8)$$
$$\text{blue}$$

The mixed precipitate of $Ni(OH)_2$ and $Fe(OH)_3$ formed by Reactions 3 and 4 is dissolved by adding a strong acid, HNO_3. An acid-base reaction occurs:

$$Ni(OH)_2(s) + 2\,H^+(aq) \rightarrow Ni^{2+}(aq) + 2\,H_2O \quad (9)$$

$$Fe(OH)_3(s) + 3\,H^+(aq) \rightarrow Fe^{3+}(aq) + 3\,H_2O \quad (10)$$

At this point, the Ni^{2+} and Fe^{3+} ions are separated by adding ammonia. The Ni^{2+} ion is converted to the deep-blue complex $Ni(NH_3)_6^{2+}$, which stays in solution:

$$Ni^{2+}(aq) + 6\,NH_3(aq) \rightarrow Ni(NH_3)_6^{2+}(aq) \quad (11)$$
$$\text{blue}$$

while the Fe^{3+} ion, which does not readily form a complex with NH_3, is reprecipitated as $Fe(OH)_3$:

$$3\,NH_3(aq) + 3\,H_2O \rightleftharpoons 3\,NH_4^+(aq) + 3\,OH^-(aq)$$

$$\underline{Fe^{3+}(aq) + 3\,OH^-(aq) \rightarrow Fe(OH)_3(s)}$$
$$Fe^{3+}(aq) + 3\,NH_3(aq) + 3\,H_2O \rightarrow Fe(OH)_3(s) + 3\,NH_4^+(aq) \quad (12)$$
$$\text{red}$$

The confirmatory test for nickel in the solution is made by adding an organic reagent, dimethylglyoxime, $C_4H_8N_2O_2$. This gives a deep rose-colored precipitate with nickel:

$$Ni^{2+}(aq) + 2\,C_4H_8N_2O_2(aq) \rightarrow Ni(C_4H_7N_2O_2)_2(s) + 2\,H^+(aq) \quad (13)$$
$$\text{rose}$$

We can confirm Fe^{3+} by dissolving the precipitate of $Fe(OH)_3$ in HCl (Reaction 10) and adding KSCN solution. If iron(III) is present, the blood-red $FeSCN^{2+}$ complex ion will form:

$$Fe^{3+}(aq) + SCN^-(aq) \rightarrow FeSCN^{2+}(aq) \quad (14)$$
$$\text{red}$$

Experimental Procedure

 WEAR YOUR SAFETY GLASSES WHILE PERFORMING THIS EXPERIMENT

Step 1. If you are testing a solution from which the Group II ions have been precipitated (Experiment 34), remove the excess H_2S and excess acid by boiling the solution until the volume is reduced to about 1 mL. Remove any sulfur residue by centrifuging the solution.

If you are working on the analysis of Group III cations only, prepare a known solution containing Fe^{3+}, Al^{3+}, Cr^{3+}, and Ni^{2+} by mixing together 0.5-mL portions of each of the appropriate 0.1 M solutions containing those cations.

Step 2. Oxidation of Cr(III) to Cr(VI) and Separation of Insoluble Hydroxides. Add 1 mL 6 M NaOH to 1 mL of the known solution in a 30-mL beaker. Boil very gently for a minute, stirring to minimize bumping. Remove heat, and slowly add 1 mL 1 M NaClO, sodium hypochlorite. Swirl the beaker for 30 seconds, using your tongs if necessary. Then boil the mixture gently for a minute. Add 0.5 mL 6 M NH_3 and let stand for 30 seconds. Then boil for another minute. Transfer the mixture to a test tube and centrifuge out the solid, which contains iron and nickel hydroxides. Decant the solution, which contains chromium and aluminum (CrO_4^{2-} and $Al(OH)_4^-$) ions, into a test tube (Step 3). Wash the solid twice with 2 mL water and 0.5 mL 6 M NaOH; after mixing, centrifuge each time, discarding the wash. Add 1 mL water and 1 mL 6 M H_2SO_4 to the solid and put the test tube aside (Step 6).

Step 3.   Separation of Al from Cr. Acidify the solution from Step 2 by adding 6 M acetic acid slowly until, after stirring, the mixture is definitely acidic to litmus. If necessary, transfer the solution to a 50-mL beaker and boil it to reduce its volume to about 3 mL. Pour the solution into a test tube. Add 6 M NH_3, drop by drop, until the solution is basic to litmus, and then add 0.5 mL in excess. Stir the mixture for a minute or so to bring the system to equilibrium. If aluminum is present, a light, translucent, gelatinous white precipitate of $Al(OH)_3$ should be floating in the clear (possibly yellow) solution. Centrifuge out the solid, and transfer the liquid, which may contain CrO_4^{2-}, into a test tube (Step 5).

Step 4.   Confirmation of the Presence of Aluminum. Wash the precipitate from Step 3 with 3 mL water once or twice, while warming the test tube in the water bath and stirring well. Centrifuge and discard the wash. Dissolve the precipitate in 2 drops 6 M $HC_2H_3O_2$, acetic acid, no more, no less. Add 3 mL water and 2 drops catechol violet reagent and stir. If Al^{3+} is present, the solution will turn blue.

Step 5.   Confirmation of the Presence of Chromium. If the solution from Step 3 is yellow, chromium is probably present; if it is colorless, chromium is absent. To the solution add 0.5 mL 1 M $BaCl_2$. In the presence of chromium you obtain a finely divided yellow precipitate of $BaCrO_4$, which may be mixed with a white precipitate of $BaSO_4$. Put the test tube in the boiling water bath for a few minutes; then centrifuge out the solid and discard the liquid. Wash the solid with 2 mL water; centrifuge and discard the wash. To the solid add 0.5 mL 6 M HNO_3, and stir to dissolve the $BaCrO_4$. Add 1 mL water, stir the orange solution, and add two drops of 3% H_2O_2, hydrogen peroxide. A deep blue solution, which may fade quite rapidly, is confirmatory evidence for the presence of chromium.

Step 6.   Separation of Iron and Nickel. Returning to the precipitate from Step 2, stir to dissolve the solid in the H_2SO_4. If necessary, warm the test tube in the water bath to complete the solution process. Then add 6 M NH_3 until the solution is basic to litmus. At that point iron will precipitate as brown $Fe(OH)_3$. Add 1 mL more of the NH_3, and stir to bring the nickel into solution as $Ni(NH_3)_6^{2+}$ ion. Centrifuge and decant the liquid into a test tube. Save the precipitate (Step 8).

Step 7.   Confirmation of the Presence of Nickel. If the solution from Step 6 is blue, nickel is probably present. To that solution add 0.5 mL dimethylglyoxime reagent. Formation of a rose-red precipitate proves the presence of nickel.

Step 8.   Confirmation of the Presence of Iron. Dissolve the precipitate from Step 6 in 0.5 mL 6 M HCl. Add 2 mL water and stir. Then add 2 drops of 0.5 M KSCN. Formation of a deep red solution of $FeSCN^{2+}$ is a definitive test for iron.

Step 9. When you have completed your analysis of the known solution, obtain a Group III unknown and test it for the possible presence of Fe^{3+}, Al^{3+}, Cr^{3+}, and Ni^{2+}.

DISPOSAL OF REAGENTS. As you complete each step in the procedure, put the waste products into a beaker. When you are finished with the experiment, pour the contents of the beaker into a waste crock, unless otherwise directed by your instructor.

Name _____   *Section* _____

Observations and Report Sheet: Qualitative Analysis of Group III Cations

Flow Diagram for Group III

Observations on known (record on diagram if different from those on prepared diagram).

Observations on unknown (record on diagram in colored pencil to distinguish from observations on known).

Unknown no. _____

Ions reported present _____ _____ _____ _____

Experiment 35 *Name* ____________________ *Section* __________

Advance Study Assignment: Qualitative Analysis of Group III Cations

1. Prepare a flow diagram for the separation and identification of the ions in Group III and put it on your report sheet.

2. Write balanced net ionic equations for the following reactions:

a. Dissolving of $Fe(OH)_3$ in nitric acid.

b. Oxidation of Cr^{3+} to CrO_4^{2-} by ClO^- in alkaline solution (ClO^- is converted to Cl^-).

c. The confirmatory test for Ni^{2+}.

d. The confirmatory test for Fe^{3+}.

3. A solution may contain any of the Group III cations. Treatment of the solution with ClO^- in alkaline medium yields a yellow solution and a colored precipitate. The acidified solution is unaffected by treatment with NH_3. The colored precipitate dissolves in nitric acid; addition of excess NH_3 to this acid solution produces only a blue solution. On the basis of this information, which Group III cations are present, absent, or still in doubt?

Present __________

Absent __________

In doubt __________

Experiment 36

THE TEN TEST TUBE MYSTERY

Having carried out at least some of the previous experiments on qualitative analysis, you should have acquired some familiarity with the properties of the cations and anions you studied. Along with this you should at this point have had some experience with interpretation of the data in Appendix II on the solubilities of ionic substances in different media.

In this experiment we are going to ask you to apply what you have learned to a related but somewhat different kind of problem. You will be furnished ten numbered test tubes in each of which there will be a solution of a single substance. You will be provided, one week in advance, with a list giving the formula and molarity of each of the ten solutes that will be used. Your problem in the laboratory will be to find out which solution is in which test tube, that is, to assign a test tube number to each of the solution compositions. You are to do this by intermixing small volumes of the solutions in the test tubes. No external reagents or acid-base indicators such as litmus are allowed. You are permitted, however, to use the odor and color of the different solute species and to make use of heat effects in reactions in your system for identification. Each student will have a different set of test tube number-solution composition correlations, and there will be several different sets of compositions as well.

Of the ten solutions, four are common laboratory reagents. They are 6 M HCl, 3 M H_2SO_4, 6 M NH_3, and 6 M NaOH. The other solutions will usually be 0.1 M nitrate, chloride, or sulfate solutions of the cations which have been studied in previous experiments in this manual (Experiments 11, 33, 34, and 35).

To determine which solution is in each test tube, you will need to know what happens when the various solutions are mixed, one with another. In some cases, nothing happens that you can observe. This will often be the case when a solution containing one of the cations is mixed with a solution of another. When one of the reagents is mixed with a cation solution you may get a precipitate, white or colored, and that precipitate may dissolve in excess reagent by complex ion formation. In a few cases a gas may be evolved. When one laboratory reagent is mixed with another, you may find that the resulting solution gets very hot and/or that a visible vapor is produced.

There is no way that you will be able to solve your particular test tube mystery without doing some preliminary work. You will need to know what to expect when any two of your ten solutions are mixed. You can find this out, for any pair, by scrutiny of Appendix II, by consulting your chemistry text, and by referring to various reference works on qualitative analysis.* A convenient way to tabulate the information you obtain is to set up a matrix with ten columns and ten rows, one for each solution. The key information about a mixture of two solutions is put in the space where the row for one solution and the column for the other intersect (as is done in Appendix II for various cations and anions). In that space you might put an NR, to indicate no apparent reaction on mixing of the two solutions. If a precipitate forms, put a P, followed by a D if the precipitate dissolves in excess reagent. If the precipitate or final solution is colored, state the color. If heat is evolved, write an H; if a gas or smoke is

---

* For example, the following references may be useful:
1. *Vogel's Qualitative Inorganic Analysis.* 6th edition, by G. Svehla, Longmans Scientific and Technical (Essex), 1987.
2. *Qualitative Analysis and the Properties of Ions in Aqueous Solutions.* 2nd edition by E. J. Slowinski and W. L. Masterton, Saunders College Publishing, 1989.
3. *Handbook of Chemistry and Physics,* Chemical Rubber Publishing, 1993.

formed, put a G or S. Since mixing solution A with B is the same as mixing B with A, not all 100 spaces in the 10-by-10 matrix need to be filled. Actually, there are only forty-five possible different pairs, since A with A is not very informative either.

Because you are allowed to use the odor or color of a solution to identify it, the problem is somewhat simpler than it might first appear. In each set of ten solutions you will probably be able to identify at least two solutions by odor and color tests. Knowing those solutions, you can make mixtures with the other solutions in which one of the components is known. From the results obtained with those mixtures, and the information in the matrix, you can identify other solutions. These can be used to identify still others, until finally the entire set of ten is identified unequivocally.

Let us now go through the various steps that would be involved in the solution to a somewhat simpler problem than the one you will solve. There are several steps, including constructing the reaction matrix, identifying solutions by simple observations, and a rationale for efficient mixing tests. Let us assume we have to identify the following six solutions in numbered test tubes:

0.1 M $NiSO_4$	0.1 M $BiCl_3$ (in 3 M HCl)
0.1 M $BaCl_2$	6 M NaOH
0.1 M $Al(NO_3)_3$	3 M H_2SO_4

Construction of Reaction Matrix

In the matrix there will be six rows and six columns, one for each of the solutions. The matrix would be set up as in Table 36.1.

Table 36.1 Reaction Matrix for Six Solutions

	$Al(NO_3)_3$	$BaCl_2$	$NiSO_4$	$BiCl_3$	$NaOH$	H_2SO_4
$Al(NO_3)_3$		NR	NR	NR	P, D	NR
$BaCl_2$			P	NR	sl P	P
$NiSO_4$ (green)				NR	P (green)	NR
$BiCl_3$					P, H	NR
NaOH						H
H_2SO_4						

Each solution contains a cation and an anion, both of which need to be considered. We need to include every possible pair of solutions; for six solutions there are 15 possible pairs. So we need 15 pieces of information as to what happens when a pair of solutions is mixed. The first pair we might consider is $Al(NO_3)_3$ and $BaCl_2$, containing Al^{3+}, NO_3^-, Ba^{2+}, and Cl^- ions. Consulting Appendix II, we see that in the Al^{3+}, Cl^- space there is an S, meaning that $AlCl_3$ is soluble and will not precipitate when aluminum and chloride ions are mixed. Similarly, $Ba(NO_3)_2$ is soluble because all nitrates are soluble. This means that there will be no reaction when the solutions of $Al(NO_3)_3$ and $BaCl_2$ are mixed. So, in the space for $Al(NO_3)_3$-$BaCl_2$, we have written NR. The same is true if we mix solutions of $Al(NO_3)_3$ and $NiSO_4$, so we insert NR in that space. Since $BiCl_3$ won't react with $Al(NO_3)_3$ solutions, there is an NR in that space too. However, when $Al(NO_3)_3$ is mixed with NaOH, one of the possible products is $Al(OH)_3$, which we see in Appendix II is insoluble in water, but dissolves in acid and excess OH^- ion (A, B). This means that on adding NaOH to $Al(NO_3)_3$, we would initially get a precipitate of $Al(OH)_3$, P, but that it would dissolve in excess NaOH, D. The precipitate is white and the solution is colorless, so we have just a P and a D in the space for that mixture.

Proceeding now to the $BaCl_2$ row, we don't need to consider the first two columns, because they would give no new information. If $BaCl_2$ and $NiSO_4$ were mixed, $BaSO_4$ is a possible product. In Appendix II we see that $BaSO_4$ is insoluble in all common solvents (I). Hence, on mixing those solutions, we would get a precipitate of $BaSO_4$; P. Using the *Handbook of Chemistry and Physics,* or some other source, we find that it is white, so a P goes into that space. With $BiCl_3$ there would be no reaction; with NaOH we might get a slight precipitate (S^- for $Ba(OH)_2$ in Appendix II), so we put a P in the appropriate space. With H_2SO_4, we would again expect to get a white precipitate of $BaSO_4$, P.

In the row for $NiSO_4$ we note that the color of the solution is green, since in Appendix II we see that $Ni^{2+}(aq)$ ion is green. The first column is that for $BiCl_3$; no precipitate forms on mixing solutions of $NiSO_4$ and $BiCl_3$, so NR is in that space. With NaOH, $Ni(OH)_2$ would precipitate, since the entry in Appendix II for Ni^{2+}, OH^-, is not S. From the *Handbook* or other sources we find that the precipitate is green, so in the space we put a P (green). In Appendix II the $Ni(OH)_2$ entry is A, N, which tells us that the precipitate will dissolve in 6 M NH_3 (but not in 6 M NaOH since there is no B in the space). There would be no reaction between solutions of $NiSO_4$ and H_2SO_4(NR).

With $BiCl_3$ in 3 M HCl we have both a salt and an acid in the solution. Addition of 6 M NaOH should produce a precipitate (white) of $Bi(OH)_3$, P. There would also be a substantial exothermic acid-base reaction between H^+ ions in the HCl and OH^- ions in the NaOH, so there would be a very noticeable rise in temperature on mixing, as denoted by H in the space. There would be no reaction with H_2SO_4.

When 6 M NaOH is mixed with 3 M H_2SO_4 there should be a large heat effect because of the acid-base reaction that occurs, so in the space for NaOH-H_2SO_4 we have an H.

The matrix in Table 36.1 summarizes the reaction information that will be needed for identification of the solutions in the test tubes. Your matrix can be constructed by the reasoning used in making this one.

Identifying Solutions by Simple Observations

When we made the matrix for the six test tube system, we noted that the $NiSO_4$ solution would be green, since Ni^{2+} ion is green. So, as soon as we see the six test tubes, we can pick out the 0.1 M $NiSO_4$. Let us assume that it is in Test Tube 4. We have now identified one solution out of six.

Selection of Efficient Mixing Tests

Knowing that Test Tube 4 contains $NiSO_4$, we mix the solution in Test Tube 4 with all of the other solutions. For each mixture, we record what we observe. The results might be as follows:

4 + 1   no obvious reaction

4 + 2   no obvious reaction

4 + 3   green precipitate forms

4 + 5   no obvious reaction

4 + 6   white precipitate

Referring to the matrix, we would expect precipitates for mixtures of $NiSO_4$ and NaOH, and $NiSO_4$ and $BaCl_2$. The former precipitate would be green, the latter white. Clearly the solution in Test Tube 3 is 6 M NaOH. The solution in Test Tube 6 must be 0.1 M $BaCl_2$. At this point three solutions have been identified.

Because 6 M NaOH reacts with several of the solutions, and because we know now that it is in Test Tube 3, we mix the solution in Test Tube 3 with those that remain unidentified, with the following results:

3 + 1   solution becomes hot

3 + 2   white precipitate, solution gets hot

3 + 5   apparently no reaction

Again, referring to the matrix we would expect heat evolution with mixtures of NaOH and H_2SO_4, and NaOH and the HCl in the $BiCl_3$. In addition, we would get a white precipitate with $BiCl_3$. It is obvious that solution 1 is 3 M H_2SO_4. Solution 2 must be 0.1 M $BiCl_3$ in 3 M HCl. By elimination solution 5 must be $Al(NO_3)_3$. We repeat the 3 + 5 mixture, since there should have been an initial precipitate that dissolved in excess NaOH. This time, using one drop of Solution 3 added to 1 mL of Solution 5, we get a precipitate. In excess of Solution 3 the precipitate dissolves as the aluminum hydroxide complex ion is formed.

The approach we have used in our example should be effective for identification of the solutions in your set. After you have tentatively assigned each solution to a test tube, carry out some confirmatory mixing reactions, so that for each solution you have at least two mixtures that give results that support your conclusions.

Wear your safety glasses while performing this experiment.

Dispose of all reaction products in the waste crock, unless directed otherwise by your instructor.

Data and Observations: Ten Test Tube Mystery

Solutions That Were Identified Directly

Solution	*Observations*	*Conclusions*	*Test tube no.*

Mixing Tests: First Series

Test tube nos.		*Observations*	*Conclusions*

Mixing Tests: Second Series

Test tube nos.		*Observations*	*Conclusions*

Confirming Tests (total of two per solution)

Test tube nos.		*Observations*	*Conclusions*

Final identifications: No. 1 _____   No. 2 _____   No. 3 _____

No. 4 _____   No. 5 _____   No. 6 _____   No. 7 _____

No. 8 _____   No. 9 _____   No. 10 _____

Unknown set no. _____

Advance Study Assignment: Ten Test Tube Mystery

1. Construct the reaction matrix for your set of ten solutions. This should go on the reverse side of this page.

2. Which solutions should you be able to identify by simple observations?

3. Outline the procedure you will follow in identifying the solutions that will require mixing tests. Be as specific as you can about what you will look for and what conclusions you will be able to draw from your observations.

LABORATORY EXAMINATION ON QUALITATIVE ANALYSIS OF CATIONS

In Experiments 33, 34, and 35 you worked with a procedure for the qualitative analysis of 11 common cations. By successive application of the procedures used for Groups I, II, and III, you could readily analyze a general unknown for the presence of the eleven ions.

Rather than give you a general unknown, we will give you a more limited unknown, one confined to five possible ions from the group of eleven. Each student will be given a different set of ions to work with and will be assigned the set of ions the week previous to this examination; that week he or she will be given a set of five letters, corresponding to the five cations that may be in the unknown, according to the following code:

$$A = Ag^+ \quad C = Hg_2^{2+} \quad E = Bi^{3+} \quad G = Sb^{3+} \quad I = Al^{3+} \quad K = Fe^{3+}$$
$$B = Pb^{2+} \quad D = Cu^{2+} \quad F = Sn^{4+} \quad H = Ni^{2+} \quad J = Cr^{3+}$$

Before coming to the laboratory for the examination, prepare a flow diagram for the analysis of your five ions. During the examination you will analyze your unknown according to your analysis scheme and will report your results on the page with your flow diagram.

You will be allowed 50 minutes in the laboratory period to complete the analysis. You will receive a bonus if you finish your unknown within 30 minutes and a penalty if it takes you more than 50 minutes. No results will be accepted after 75 minutes.

Your grade on the examination will be based on (1) the accuracy of your analysis, (2) the workability of the scheme for analysis on your flow diagram, (3) the absence of extra steps in your flow sheet, steps that relate to ions other than those that can be in your unknown, and (4) the time it takes you to complete the analysis.

There is to be no communication between students during the examination. During the laboratory period, your laboratory supervisor will not answer any questions concerning the analysis. In preparing your flow diagram you may use the portions of the procedure for Groups I, II, and III, which pertain to the cations that may be in your unknown, or you may use other reactions that you are sure will enable you to make the analysis. It is to your interest to shorten the scheme as much as you can, avoiding any steps that are unnecessary. It is of course important that the scheme you decide on is one that will work for the set of ions in your unknown.

In developing your analysis scheme there are several precautions you should observe, which are perhaps not at once obvious. Your procedure will in all probability involve removing from the solution first one group of ions, then another, and then perhaps a third. This means that if you are to avoid difficulty in interpretation, the group separation must be *complete*: Always test to make sure that the precipitating reagent has been added in sufficient amount to bring down all of the material that should precipitate at that point. If all of the Group I ions are not removed by the addition of an adequate amount of HCl, you may be sure that the remaining Group I ions will precipitate along with Group II when that group is removed from the system.

It is also essential to recognize that when groups of ions are separated by means of a selective precipitation, decantation after centrifuging will leave some of the solution along with the solid in the test tube. The Group I precipitate may be washed with a few drops of cold water to remove ions in Groups II and III. The Group II precipitate may be washed with 1 M NH_4Cl solution to remove Group III cations. The wash liquid should be thoroughly mixed with the solid to dilute the undesired cations. Centrifuge, decant, and discard the wash liquid.

Lead chloride is relatively soluble, so that although most of the Pb^{2+} will be removed in Group I, some PbS may precipitate in Group II. It will be dissolved along with the Bi^{3+} and Cu^{2+} in nitric acid, and will precipitate along with $Bi(OH)_3$. It will not, however, interfere with the test for Bi^{3+} that we have been using.

Because in this experiment you may be carrying out some reactions with which you are not familiar, use due caution and **remember to wear your safety glasses.**

DISPOSAL OF REAGENTS. As you go through your procedure, pour the reaction products into a beaker. At the end of the experiment, pour the contents of the beaker into the waste crock, unless directed otherwise by your instructor.

Experiment 37   *Name* _____   *Section* _____

Advance Study Assignment and
Laboratory Report: Laboratory Examination

Unknown no. _____

Possible ions. _____   _____   _____   _____   _____

Flow Diagram

Ions found in unknown _____   _____   _____   _____   _____

Time turned in (to be entered by laboratory supervisor)   _____

Experiment 38

SOME NONMETALS AND THEIR COMPOUNDS—PREPARATIONS AND PROPERTIES

Some of the most commonly encountered chemical substances are nonmetallic elements or their simple compounds. O_2 and N_2 in the air, CO_2 produced by combustion of oil, coal, or wood, and H_2O in rivers, lakes, and air are typical of such substances. Substances containing nonmetallic atoms, whether elementary or compound, are all molecular, reflecting the covalent bonding that holds their atoms together. They are often gases, due to weak intermolecular forces. However, with high molecular masses, hydrogen bonding, or macromolecular structures one finds liquids, like H_2O and Br_2, and solids, like I_2 and graphite.

Several of the common nonmetallic elements and some of their gaseous compounds can be prepared by simple reactions. In this experiment you will prepare some typical examples of such substances and examine a few of their characteristic properties.

Experimental Procedure

WEAR YOUR SAFETY GLASSES WHILE PERFORMING THIS EXPERIMENT

In several of the experiments you will be doing you will prepare gases. In general we will not describe in detail what you should observe, so perform each preparation carefully and report what you actually observe, not what you think we expect you to observe.

There are several tests we will make on the gases you prepare, and the way each of these should be carried out is summarized below.

TEST FOR ODOR. To determine the odor of a gas, first pass your hand across the end of the tube, bringing the gas toward your nose. If you don't detect an odor, sniff near the end of the tube, first at some distance and then gradually closer. Don't just put the tube at the end of your nose immediately and take a deep breath. Some of the gases you will make have no odor, and some will have very impressive ones. Some of the gases are very toxic and, even though we will be making only small amounts, caution in testing for odor is important.

TEST FOR SUPPORT OF COMBUSTION. A few gases will support combustion, but most will not. To make the test, ignite a wood splint with a Bunsen flame, blow out the flame, and put the glowing, but not burning, splint into the gas in the test tube. If the gas supports combustion, the splint will glow more brightly, or may make a small popping noise. If the gas does not support combustion, the splint will go out almost instantly. You can use the same splint for all the support of combustion tests.

TEST FOR ACID-BASE PROPERTIES. Many gases are acids. This means that if the gas is dissolved in water it will produce some H^+ ions. A few gases are bases; in water solution such gases produce OH^- ions. Other gases do not interact with water and are neutral. It is easy to establish the acidic or basic nature of a gas by using a chemical indicator. One of the most common acid-base indicators is litmus, which is red in acidic solution and blue in basic solution. To test whether a gas is an acid, moisten a piece of blue litmus paper with water from your wash bottle, and put the paper down in the test tube in which the gas is present. If the gas is an acid, the paper will turn red. Similarly, to test if the gas is a base, moisten a piece of red litmus paper, and hold it down in the tube. A color change to blue will occur if the gas forms a basic solution. Since you may have used an acid or a base in making the gas, do not touch the walls of the test tube with the paper. The color change will occur fairly quickly and smoothly over the surface of the paper if the gas is acidic or basic. It is not necessary to use a new piece of litmus for each test. Start with a piece of blue and a piece of red litmus. If you need to regenerate the blue paper, hold it over an open bottle of 6 M NH_3, whose vapor is basic. If you need to make red litmus, hold the moist paper over an open bottle of 6 M acetic acid.

A. Preparation and Properties of Nonmetallic Elements: O_2, N_2, Br_2, I_2

1. OXYGEN, O_2. Oxygen can be easily prepared in the laboratory by the decomposition of H_2O_2, hydrogen peroxide, in aqueous solution. Hydrogen peroxide is not very stable and will break down to water and oxygen gas on addition of a suitable catalyst, particularly MnO_2:

$$2\ H_2O_2(aq) \xrightarrow{MnO_2} O_2(g) + 2\ H_2O \qquad (1)$$

Add 1 mL 3% H_2O_2 solution in water to a small test tube. Pick up a small amount (~ 0.1 g) of MnO_2 on the tip of your spatula and add it to the liquid in the tube. Hold your finger over the end of the tube to help confine the O_2. Test the evolved gas for odor. Test the gas for any acid-base properties, using moist blue and red litmus paper. Test the gas for support of combustion. Record your observations.

2. NITROGEN, N_2. Sodium sulfamate in water solution will react with nitrite ion to produce nitrogen gas:

$$NO_2^-(aq) + NH_2SO_3^-(aq) \rightarrow N_2(g) + SO_4^{2-}(aq) + H_2O \qquad (2)$$

To a small test tube add about 1 mL 1 M KNO_2, potassium nitrite. Add 10 to 12 drops 0.5 M $NaNH_2SO_3$, sodium sulfamate, and place the test tube in a hot-water bath made from a 250-mL beaker half full of water. Bubbles of nitrogen should form within a few moments. Confine the gas for a few seconds with a stopper. Cautiously test the evolved gas for odor. Carry out the tests for acid-base properties and for support of combustion. If you need to generate more N_2 to complete the tests, add sodium sulfamate solution as necessary, 5 to 10 drops at a time. Report your observations.

3. IODINE, I_2. The halogen elements are most readily prepared from their sodium or potassium salts. The reaction involves an oxidizing agent, which can remove electrons from the halide ions, freeing the halogen. The reaction that occurs when a solution of potassium iodide, KI, is treated with 6 M HCl and a little MnO_2 is

$$2\ I^-(aq) + 4\ H^+(aq) + MnO_2(s) \rightarrow I_2(aq) + Mn^{2+}(aq) + 2\ H_2O \qquad (3)$$

At 25°C I_2 is a solid, with relatively low solubility in water, and appreciable volatility. I_2 can be extracted from aqueous solution into organic solvents, particularly hexane, C_6H_{14} (HEX). The solid, vapor, and solutions in water and HEX all have characteristic colors.

Put two drops 1 M KI into a *regular* (18 × 150 mm) test tube. Add six drops 6 M HCl and a tiny amount of manganese dioxide, MnO_2. Swirl the mixture and note any changes that occur. Put the test tube into a hot-water bath. After a minute or two a noticeable amount of I_2 vapor should be visible above the liquid. Remove the test tube from the water bath and add 10 mL distilled water. Stopper the tube and shake. Note the color of I_2 in the solution. Sniff the vapor above the solution. Decant the liquid into another regular test tube and add 3 mL hexane, C_6H_{14}. Stopper and shake the tube. Observe the color of the HEX layer and the relative solubility of I_2 in water and HEX. Record your observations.

4. BROMINE, Br₂. Bromine can be made by the same reaction as is used to make iodine, substituting bromide ion for iodide. Bromide ion is less easily oxidized than iodide ion. Bromine at 25°C is a liquid. Br_2 can be extracted from water solution into hexane. Its liquid, vapor, and solutions in water and HEX are colored.

To two or three drops 1 M NaBr in a regular test tube add six drops 6 M HCl and a small amount of MnO_2, about the size of a small pea. Swirl to mix the reagents, and observe any changes. Heat the tube in the water bath for a minute or two. Try to detect Br_2 vapor above the liquid by observing its color. Add 10 mL water to the tube, stopper, and shake. Note the color of Br_2 in the solution. Sniff the vapor, cautiously. Decant the liquid into a regular test tube. Add 3 mL HEX, stopper, and shake. Note the color of Br_2 in HEX. Record your observations.

B. Preparation and Properties of Some Nonmetallic Oxides: CO₂, SO₂, NO, and NO₂

1. CARBON DIOXIDE, CO₂. Carbon dioxide, like several nonmetallic oxides, is easily made by treating an oxyanion with an acid. With carbon dioxide the oxyanion is CO_3^{2-}, carbonate ion, which is present in solutions of carbonate salts, such as Na_2CO_3. The reaction is

$$CO_3^{2-}(aq) + 2\ H^+(aq) \rightarrow H_2CO_3(aq) \rightarrow CO_2(g) + H_2O \tag{4}$$

Carbon dioxide is not very soluble in water, and on acidification, carbonate solutions will tend to effervesce as CO_2 is liberated. Nonmetallic oxides in solution are often acidic but never basic.

To 1 mL M Na_2CO_3 in a small test tube add six drops 3 M H_2SO_4. Test the gas for odor, acidic properties, and ability to support combustion.

2. SULFUR DIOXIDE, SO₂. Sulfur dioxide is readily prepared by acidification of a solution of sodium sulfite, containing sulfite ion, SO_3^{2-}. As you can see, the reaction that occurs is very similar to that with carbonate ion.

$$SO_3^{2-}(aq) + 2\ H^+(aq) \rightarrow H_2SO_3(aq) \rightarrow SO_2(g) + H_2O \tag{5}$$

Sulfur dioxide is considerably more soluble in water than is carbon dioxide. Some effervescence may be observed on acidification of concentrated sulfite solutions, which increases if the solution is heated in a water bath.

To 1 mL M Na_2SO_3 in a small test tube add six drops 3 M H_2SO_4. *Cautiously* test the evolved gas for odor. Test its acidic properties and its ability to support combustion. Put the test tube into the water bath for a few seconds to see if effervescence occurs if the solution is hot.

3. NITROGEN DIOXIDE, NO₂, AND NITRIC OXIDE, NO. If a solution containing nitrite ion, NO_2^-, is treated with acid, two oxides are produced, NO_2 and NO. In solution these gases are combined in the form of N_2O_3, which is colored. When these gases come out of solution, the mixture contains NO and NO_2; the latter is colored. NO is colorless and reacts readily with oxygen in the air to form NO_2. The preparation reaction is

$$2\ NO_2^-(aq) + 2\ H^+(aq) \rightarrow N_2O_3(aq) + H_2O \rightarrow NO(g) + NO_2(g) + H_2O \tag{6}$$

To 1 mL of 1 M KNO_2 in a small test tube, add six drops 3 M H_2SO_4. Swirl the mixture for a few seconds and note the color of the solution. Warm the tube in the water bath for a few seconds to increase the rate of gas evolution. Note the color of the gas that is given off. Cautiously test the odor of the gas. Test its acidic properties and its ability to support combustion. Record your observations.

C. Preparation and Properties of Some Nonmetallic Hydrides: NH_3, H_2S

1. AMMONIA, NH_3. A 6-M solution of NH_3 in water is a common laboratory reagent. You may have been using it in this experiment to make your litmus paper turn blue. Ammonia gas can be made by simply heating 6 M NH_3. It can also be prepared by addition of a strongly basic solution to a solution of an ammonium salt, such as NH_4Cl. The latter solution contains NH_4^+ ion. On treatment with OH^- ion, as in a solution of NaOH, the following reaction occurs:

$$NH_4^+(aq) + OH^-(aq) \rightarrow NH_3(aq) + H_2O \rightarrow NH_3(g) + H_2O \qquad (7)$$

The odor of NH_3 is characteristic. NH_3 is very soluble in water, so effervescence is not observed, even on heating concentrated solutions.

Add 1 mL 1 M NH_4Cl to a small test tube. Add 1 mL 6 M NaOH. Swirl the mixture and test the odor of the evolved gas. Put the test tube into the water bath for a few moments to increase the amount of NH_3 in the gas phase. Test the gas with moistened blue and red litmus paper. Test the gas for support of combustion.

2. HYDROGEN SULFIDE, H_2S. Hydrogen sulfide can be made by treating some solid sulfides, particularly FeS, with an acid such as HCl or H_2SO_4. With FeS the reaction is

$$FeS(s) + 2 H^+(aq) \rightarrow H_2S(g) + Fe^{2+}(aq) \qquad (8)$$

This reaction was used for many years to make H_2S in the laboratory in courses in qualitative analysis. In this experiment we will employ the method currently used in such courses for H_2S generation. This involves the decomposition of thioacetamide, CH_3CSNH_2, which occurs in solution on treatment with acid and heat. The reaction is

$$CH_3CSNH_2(aq) + 2 H_2O \rightarrow H_2S(g) + CH_3COO^-(aq) + NH_4^+(aq) \qquad (9)$$

The odor of H_2S is notorious. H_2S is moderately soluble in water and, since it is produced reasonably slowly in Reaction 9, there will be little if any effervescence.

To 1 mL of 1 M thioacetamide in a small test tube add six drops 3 M H_2SO_4. Put the test tube in a boiling-water bath for about 1 minute. There may be some cloudiness due to formation of free sulfur. Carefully smell the gas in the tube; H_2S is toxic. Test the gas for acidic and basic properties and for the ability to support combustion.

D. Identification of Unknown Solution (Optional)

In this experiment you prepared nine different species containing nonmetallic elements. In each case the source of the species was in solution. In this part of the experiment we will give you an unknown solution that can be used to make one of the nine species. It will be a solution used in preparing one of the species, but it will not be an acid or a base. Identify by suitable tests the species that can be made from your unknown and the substance that is present in your unknown solution.

CAUTION: *The following gases prepared in this experiment are toxic: Br_2, SO_2, NO_2, NH_3, and H_2S. Do not inhale these gases unnecessarily when testing their odors. One small sniff will be sufficient and will not be harmful. It is good for you to know these odors, in case you encounter them in the future.*

DISPOSAL OF REACTION PRODUCTS. Most of the chemicals used in this experiment can be discarded down the sink drain. Pour the solutions from Sections 3 and 4 of Part A, containing I_2 and Br_2 in hexane, into a waste crock, unless directed otherwise by your instructor.

Name _____   *Section* _____

Data and Observations: Nonmetals and Their Compounds

A.

	Element Prepared	Degree of Effervescence	Odor	Acid-Base Tests	Support of Combustion Test
1.	O_2	_____	_____	_____	_____
2.	N_2	_____	_____	_____	_____

		Odor	Vapor	Color In H₂O	In HEX
				Color	
		Odor	Vapor	In H_2O	In HEX
3.	I_2	_____	_____	_____	_____
4.	Br_2	_____	_____	_____	_____

B.

	Oxide Prepared	Degree of Effervescence	Odor	Acid Test	Color	Support of Combustion Test
1.	CO_2	_____	_____	_____	_____	_____
2.	SO_2	_____	_____	_____	_____	_____
3.	$NO_2 + NO$	_____	_____	_____	soln _____	_____
					gas _____	

C.

	Hydride Prepared	Degree of Effervescence	Odor	Acid-Base Tests	Support of Combustion Test
1.	NH_3	_____	_____	_____	_____
2.	H_2S	_____	_____	_____	_____

D. Properties of Unknown (Optional)

Nonmetal species that can be made from unknown _____

Identity of unknown solution _____

Unknown no. _____

Advance Study Assignment: Nonmetals and Their Compounds

In this experiment nine species are prepared and studied. For each of the species, list the reagents that are used in its preparation and the reaction that occurs.

	Reagents Used	*Reaction*
1. O_2		
2. N_2		
3. I_2		
4. Br_2		
5. CO_2		
6. SO_2		
7. $NO_2 + NO$		
8. NH_3		
9. H_2S		

Experiment 39

SPOT TESTS FOR SOME COMMON ANIONS

There are two broad categories of problems in analytical chemistry. Quantitative analysis deals with the determination of the amounts of certain species present in a sample; there are several experiments in this manual involving quantitative analysis, and you probably have performed some of them. The other area of analysis, called qualitative analysis, has a more limited purpose, establishing whether given species are or are not present in detectable amounts in a sample. Several of the experiments in this manual have dealt with problems in qualitative analysis.

One can carry out the qualitative analysis of a sample in various ways. Probably the simplest approach, which we will use in this experiment, is to test for the presence of each possible component by adding a reagent that will cause the component, if it is in the sample, to react in a characteristic way. This method involves a series of ''spot'' tests, one for each component, carried out on separate samples of the unknown. The difficulty with this way of doing qualitative analysis is that frequently, particularly in complex mixtures, one species may interfere with the analytical test for another. Although interferences are common, there are many ions that can, under optimum conditions at least, be identified in mixtures by simple spot tests.

In this experiment we will use spot tests for the analysis of a mixture that may contain the following commonly encountered ions in solution:

$$CO_3^{2-} \quad PO_4^{3-} \quad Cl^- \quad SCN^-$$

$$SO_4^{2-} \quad SO_3^{2-} \quad C_2H_3O_2^- \quad NO_3^-$$

The procedures we will use involve simple acid-base, precipitation, complex ion formation, or oxidation-reduction reactions. In each case you should try to recognize the kind of reaction that occurs, so that you can write the net ionic equation that describes it.

Experimental Procedure

WEAR YOUR SAFETY GLASSES WHILE PERFORMING THIS EXPERIMENT

Carry out the test for each of the anions as directed. Repeat each test using a solution made by diluting the anion solution 9:1 with distilled water; use your 10-mL graduated cylinder to make the dilution and make sure you mix well before taking the sample for analysis. In some of the tests, a boiling-water bath containing about 100 mL water in a 150-mL beaker will be needed, so set that up before proceeding. When performing a test, if no reaction is immediately apparent, stir the mixture with your stirring rod to mix the reagents. These tests can easily be used to detect the anions at concentrations of 0.02 M or greater, but in dilute solutions, careful observation may be required.

Test for the Presence of Carbonate Ion, CO_3^{2-}

Cautiously add 1 mL of 6 M HCl to 1 mL of 1 M Na_2CO_3 in a small test tube. With concentrated solutions, bubbles of carbon dioxide gas are immediately evolved. With dilute solutions, the effervescence will be much less obvious. Warming in the water bath, with stirring, will increase the amount of bubble formation. Carbon dioxide is colorless and odorless.

Test for the Presence of Sulfate Ion, SO_4^{2-}

Add 1 mL of 6 M HCl to 1 mL of 0.5 M Na_2SO_4. Add a few drops of 1 M $BaCl_2$. A white, finely divided precipitate of $BaSO_4$ indicates the presence of SO_4^{2-} ion.

Test for the Presence of Phosphate Ion, PO_4^{3-}

Add 1 mL 6 M HNO_3 to 1 mL of 0.5 M Na_2HPO_4. Then add 1 mL of 0.5 M $(NH_4)_2MoO_4$ and stir thoroughly. A yellow precipitate of ammonium phosphomolybdate, $(NH_4)_3PO_4 \cdot 12$ MoO_3, establishes the presence of phosphate. The precipitate may form slowly, particularly in the more dilute solution; if it does not appear promptly, put the test tube in the boiling-water bath for a few minutes.

Test for the Presence of Sulfite Ion, SO_3^{2-}

Sulfite ion in acid solution tends to evolve SO_2, which can be detected by its odor even at low concentrations. Sulfite ion is slowly oxidized to sulfate in moist air, so sulfite-containing solutions will usually test positive for sulfate ion.

To 1 mL 0.5 M Na_2SO_3 add 1 mL 6 M HCl, and mix with your stirring rod. Cautiously, sniff the rod to try to detect the acrid odor of SO_2, which is a good test for sulfite. If the odor is too faint to detect, put the test tube in the hot water bath for 10 seconds and sniff again.

To proceed with a chemical test, add 1 mL 1 M $BaCl_2$ to the solution in the tube. Stir, and centrifuge out any precipitate of $BaSO_4$. Decant the clear solution into a test tube and add 1 mL 3% H_2O_2, hydrogen peroxide. Stir the solution and let it stand for a few seconds. If sulfite is present, its oxidation to sulfate will cause a new precipitate of $BaSO_4$ to form.

If thiocyanate is present, it may interfere with the chemical test. In that case, add 1 mL 1 M $BaCl_2$ to 1 mL of the sample. Centrifuge out any precipitate, which will contain $BaSO_3$ if sulfite is present, but will not contain thiocyanate. Wash the solid with 2 mL water, stir, centrifuge, and discard the wash. To the precipitate add 1 mL 6 M HCl and 2 mL water, and stir. Centrifuge out any $BaSO_4$, and decant the clear liquid into a test tube. To the liquid add 1 mL 3% H_2O_2. If you have sulfite present, you will observe a new precipitate of $BaSO_4$ within a few seconds.

Test for the Presence of Thiocyanate Ion, SCN^-

Add 1 mL 6 M acetic acid, $HC_2H_3O_2$, to 0.5 M KSCN and stir. Add one or two drops 0.1 M $Fe(NO_3)_3$. A deep red coloration as a result of formation of $FeSCN^{2+}$ ion is proof for the presence of SCN^- ion.

Test for the Presence of Chloride Ion, Cl^-

Add 1 mL 6 M HNO_3 to 1 mL 0.5 M NaCl. Add two or three drops 0.1 M $AgNO_3$. A white, curdy precipitate of AgCl will form if chloride ion is present.

Several anions interfere with this test, because they too form white precipitates with $AgNO_3$ under these conditions. In this experiment only SCN^- ion will interfere. If the sample contains SCN^- ion, put 1 mL of the solution into a 30- or 50-mL beaker and add 1 mL 6 M HNO_3. Boil the solution gently until its volume is decreased by half. This will destroy most of the thiocyanate. To the solution in a small test tube add 1 mL 6 M HNO_3 and a few drops of $AgNO_3$ solution. If Cl^- is present you will get a curdy precipitate. If Cl^- is absent, you may see some cloudiness due to residual amounts of SCN^-.

Test for the Presence of Acetate Ion, C₂H₃O₂⁻

To 1 mL 0.5 M NaC₂H₃O₂ add 6 M NH₃ or 6 HNO₃ until the solution is just basic to litmus. Add one drop of 1 M BaCl₂. If a precipitate forms, add 1 mL of the BaCl₂ solution to precipitate interfering anions. Stir, centrifuge, and decant the clear liquid into a test tube. Add one drop BaCl₂ to make sure that precipitation was complete. To 1 mL of the liquid add 0.1 M KI₃, drop by drop, until the solution takes on a fairly strong rust color. Add 0.5 mL 0.1 M La(NO₃)₃ and six drops 6 M NH₃. Stir, and put the test tube in the water bath. If acetate is present, the mixture will darken to nearly black in a few minutes. The color is due to iodine adsorbed on the basic lanthanum acetate precipitate.

Test for the Presence of Nitrate Ion, NO₃⁻

To 1 mL 0.5 M NaNO₃ add 1 mL 6 M NaOH. Then add a few granules of Al metal, using your spatula, and put the test tube in the hot-water bath. In a few seconds, the Al-NaOH reaction will produce H₂ gas, which will reduce the NO₃⁻ ion to NH₃, which will come off as a gas. To detect the NH₃, hold a piece of moistened red litmus paper just above the end of the test tube. If the sample contains nitrate ion, the litmus paper will gradually turn blue, within a minute or two. Blue spots caused by effervescence are not to be confused with the blue color over all of the litmus exposed to NH₃ vapors. Cautiously sniff the vapors at the top of the tube; you may be able to detect the odor of ammonia.

If SCN⁻ is present, it will interfere with the test. In that case, first add 1 mL 1 M CuSO₄ to 1 mL of the sample and put the test tube in the water bath for a minute or two. Centrifuge out the precipitate, and decant the solution into a test tube. Add 1 mL 1 Na₂CO₃ to remove excess Cu²⁺ ion. Centrifuge out the precipitate, and decant the solution into a test tube. To 1 mL of the solution add 1 mL 6 M NaOH, and proceed, starting with the second sentence of this procedure.

When you have completed all of the tests, obtain an unknown from your laboratory supervisor, and analyze it by applying the tests to separate 1-mL portions. The unknown will contain three or four ions on the list, so your test for a given ion may be affected by the presence of others. When you think you have properly analyzed your unknown, you may, if you wish, make a "known" with the composition you found and test it to see if it behaves as your unknown did.

DISPOSAL OF REAGENTS. As you complete each test, pour the products into a beaker. When you have finished the experiment, pour the contents of the beaker into the waste crock, unless directed otherwise by your instructor.

Observations and Report Sheet: Spot Tests for Some Common Anions

Observations and Comments on Spot Tests

Ion	*Stock Solution*	*9 : 1 Dilution*	*Unknown*
CO_3^{2-}			
SO_4^{2-}			
PO_4^{3-}			
SO_3^{2-}			
Cl^-			
$C_2H_3O_2^-$			
SCN^-			
NO_3^-			

Unknown no. _____ contains _____

Advance Study Assignment: Spot Tests for Some Common Ions

1. Each of the observations listed was made on a different solution. Given the observation, state which ion studied in this experiment is present. If the test is not definitive, indicate that with a question mark.

    a. Addition of 6 M NaOH and Al to the solution produces a vapor that turns red litmus blue.
        Ion present:

    b. Addition of 6 M HCl produces a vapor with an acrid odor.
        Ion present:

    c. Addition of 6 M HNO_3 produces an effervescence.
        Ion present:

    d. Addition of 6 M HNO_3 plus 0.1 M $AgNO_3$ produces a precipitate.
        Ion present:

    e. Addition of 6 M HNO_3 plus 1 M $BaCl_2$ produces a precipitate.
        Ion present:

    f. Addition of 6 M HNO_3 plus 0.5 M $(NH_4)_2MoO_4$ produces a precipitate.
        Ion present:

2. An unknown containing one or more of the ions studied in this experiment has the following properties:

    a. No effect on addition of 6 M HNO_3.
    b. No effect on addition of 0.1 M $AgNO_3$ to solution in (a).
    c. White precipitate on addition of 1 M $BaCl_2$ to solution in (a).
    d. Yellow precipitate on addition of $(NH_4)_2MoO_4$, to solution in (a).

On the basis of this information which ions are present, which are absent, and which are in doubt?

         Present   Absent   In doubt

(continued on following page)

(continued)

3. The chemical reactions that are used in the anion spot tests in this experiment are for the most part simple precipitation or acid-base reactions. Given the information in each test procedure, try to write the net ionic equation for the key reaction in each test.

   a. CO_3^{2-}

   b. SO_4^{2-}

   c. PO_4^{3-} (Reactants are HPO_4^{2-}, NH_4^+, MoO_4^{2-}, and H^+; products are $(NH_4)_3PO_4 \cdot 12 \, MoO_3$ and H_2O; no oxidation or reduction occurs.)

   d. SO_3^{2-}

   e. SCN^-

   f. Cl^-

   g. NO_3^- (Take as reactants Al and NO_3^-; as products NH_3 and AlO_2^-; final equation also contains OH^- and H_2O as reactants.)

xperiment 40

SULFUR CHEMISTRY

Although sulfur and oxygen belong to the same family in the Periodic Table, their chemical behaviors are markedly different. In general oxygen behaves as an oxidizing agent, and in most of its compounds it has an oxidation number of -2. In sulfide compounds sulfur also has an oxidation number of -2, but the reactivity of sulfur toward elements is typically much lower than that of oxygen. More often we encounter sulfur in a positive oxidation state, with $+6$ the most common, followed by $+4$.

Sulfur burns easily in air to form SO_2, a gas with a characteristic acrid odor. Sulfur dioxide dissolves readily in water to form sulfurous acid, H_2SO_3, which is a weak acid containing the sulfite anion, SO_3^{2-}. Both SO_2 and the sulfite ion are subject to further oxidation, to SO_3 and SO_4^{2-}. Sulfur trioxide reacts with the water to form sulfuric acid, H_2SO_4. This reaction will occur in your lungs if you breathe some SO_3, which accounts for the fact that sulfur trioxide is a choking gas. H_2SO_3 and H_2SO_4 are the main components in much of the acid rain that falls downwind from industrial plants that burn a lot of coal. Sulfur compounds often have bad smells, and in this experiment we will try to keep the odors down, although you may get a whiff now and then.

One property of sulfur that is not shared with oxygen is its tendency to bond to itself in chains. Elemental sulfur is crystalline and contains eight-membered rings, so its formula is S_8. Elemental sulfur will add to sulfide ion and in solution, forming polysulfides such as S_5^{2-} and S_6^{2-}. This tendency to form chains also occurs with some sulfur oxyanions.

Most sulfates are soluble in water, the most common exceptions being $BaSO_4$ and $PbSO_4$. Sulfites are typically less soluble than sulfates. $CuSO_3$, for example, is insoluble, while the sulfate is soluble. Since SO_3^{2-} ion is the conjugate base of the weak acid HSO_3^-, water-insoluble sulfites dissolve in solutions of strong acids. Insoluble sulfates, on the other hand, do not go into solution in acids, since the HSO_4^- ion is a fairly strong acid. On the stockroom shelf we find hydrogen sulfites and hydrogen sulfates, such as $NaHSO_3$ and $NaHSO_4$, as well as sulfite and sulfate salts, such as Na_2SO_3 and Na_2SO_4. Sodium hydrogen sulfate is a convenient solid source of H^+ ion.

There are many sulfur oxyanions, and in this experiment we will work with several of them. One of the most important of these is the thiosulfate ion, $S_2O_3^{2-}$, which we will make by reacting sulfur with sulfite ion in solution. Thiosulfate ion has the electronic structure of SO_4^{2-} ion, with one oxygen replaced by sulfur. Thiosulfate ion has some interesting chemistry. It forms many stable complex ions, probably the most important being that with silver ion, $Ag(S_2O_3)_2^{3-}$. In photography, silver bromide is removed from the negative in the fixing reaction; in that reaction, sodium thiosulfate, "hypo," in solution is used to dissolve the $AgBr$, with the silver ion ending up as the thiosulfate complex. Thiosulfate ion is also a reasonably good reducing agent and is used in the classic redox titration for determination of iodine:

$$2\ S_2O_3^{2-}(aq) + I_2(aq) \rightarrow 2\ I^-(aq) + S_4O_6^{2-}(aq)$$

The tetrathionate ion, a product in the above reaction, contains a chain of four sulfur atoms and is formed when two $S_2O_3^{2-}$ ions link together.

Thiosulfate ion is stable in neutral or basic solution, but in acid it will decompose to free sulfur and sulfite ion. Sodium thiosulfate is produced as a hydrate, $Na_2S_2O_3 \cdot 5\ H_2O$; that is the thiosulfate you will find in the stockroom.

In the second part of the experiment we will compare the properties of thiosulfate ion with those of sulfite and sulfate ions, with respect to reaction with acids, solubility of salts, redox behavior, and tendency to form complex ions. We will then use those properties to confirm that the salt we prepared is indeed a thiosulfate.

Experimental Procedure

A. Preparation of Sodium Thiosulfate

Weigh out about 6.0 grams of sodium sulfite in a 100-mL beaker, using a top-loading balance. Add 30-mL distilled water, and note the level of the liquid on the beaker wall. Place the beaker on a piece of wire gauze on a ring stand, and heat gently until all of the solid has dissolved (do not boil). Add about 2.0 grams of powdered sulfur, and boil the mixture gently for 10 minutes, while stirring frequently. You will find that you have to push the sulfur down the walls of the beaker occasionally during this period since the solid tends to creep up the sides.

Set up a small Buchner funnel, using a small filter flask, 250 mL or less. Filter the mixture through filter paper while hot, using suction. Rinse out the beaker with a minimal amount of water (2 or 3 mL), and pour the rinse into the funnel. Remove the filter paper and residual sulfur, and put the paper aside for weighing.

Clean out the 100-mL beaker and into it pour the filtrate from the filter flask, rinsing the flask with a few milliliters of water and adding the rinse to the beaker. Evaporate the solution by gently boiling until its volume goes down to about one half that of the initial 30 mL of solution. Leave your stirring rod in the beaker during this process. While the evaporation is going on, you may proceed with the rest of the experiment, but check the level in the beaker every few minutes.

When the liquid level has been reduced to about 15 mL, remove the beaker from the ring and put it in an ice-bath. Cool the mixture to 3°C or lower, while stirring. Scratching the bottom of the beaker with the stirring rod will tend to encourage crystal formation. If after 5 minutes at the low temperature, no crystals have appeared, seed the mixture with a few finely ground crystals of $Na_2S_2O_3 \cdot 5\ H_2O$. Once the solute ions have been given a clue as to the structure of the crystals that should form, crystallization should proceed smoothly.

Set up the small Buchner funnel, and after 5 minutes of cooling and stirring the crystals, filter them through filter paper with suction. Scrape out as many crystals from the beaker into the funnel as you can, using your rubber policeman. Do not wash the crystals, since they are very soluble in water. Pull air through the crystals for 5 minutes, while pressing them down with the policeman. Remove the filter paper from the funnel, and transfer the crystals to a dry piece of filter paper for further drying while you continue with the rest of the experiment.

B. Properties of Some Sulfur Oxyanions

1. BEHAVIOR ON ADDITION OF ACID. To three small test tubes add 1 mL of 0.1 M solutions of $Na_2S_2O_3$, Na_2SO_3, and Na_2SO_4, one solution to a tube. To the first solution add 6 drops of 6 M HCl, and wiggle the tube to mix the reagents. Observe the tube for a minute or two. Record any changes. Carefully sniff the top of the tube to detect any odor of evolved gas. Put the test tube in a hot-water bath. Test for evolved gas with a piece of moistened blue litmus paper held over the end of the tube; repeat the test with a small strip of filter paper moistened with 0.1 M $K_2Cr_2O_7$. If SO_2 is given off, the litmus paper will gradually turn red, and the orange dichromate will slowly turn green as it oxidizes any SO_2. Repeat the procedure with the tubes containing Na_2SO_3 and Na_2SO_4. Record all observations.

2. SOLUBILITY OF SALTS. Pour the contents of the test tubes into a beaker and rinse out the tubes. To the three tubes add 1 mL of the thiosulfate, sulfite, and sulfate solutions as before. Then add 6 drops 1 M $BaCl_2$ to each tube. If a precipitate forms, centrifuge it out, discarding the liquid into the waste beaker. To each precipitate add 1 mL 6 M HCl and stir with your stirring rod. Record your observations.

3. REDOX BEHAVIOR. Again, pour the contents of the tubes into the beaker, and, after rinsing them out, put 1 mL of the three sulfur oxyanion-containing solutions into the tubes, one solution to a tube. This time add 0.1 M KI_3 solution (which contains I_2 in KI), drop by drop, to each tube. If the sulfur oxyanion reduces I_2, the color will disappear. Record your observations.

4. COMPLEX ION FORMATION. Pour the contents of the tubes into the waste beaker, and, after rinsing, put 4 drops 0.1 M $AgNO_3$ into each tube. To the first tube add 1 mL 0.1 M $Na_2S_2O_3$ dropwise, taking no more than 10 seconds to add the solution. If a precipitate forms and then dissolves in excess reagent, the oxyanion must form a complex ion with Ag^+. If the solution is added too slowly, reduction of Ag^+ ion to the metal may occur, which will cause the solution to turn yellow, or even black. Repeat the experiment using the Na_2SO_3 and the Na_2SO_4 solutions. Record your observations.

To establish relative stabilities of any complex ions, we can proceed as follows. Pour out the contents of the test tubes into the beaker, rinse the tubes, and to each of the three tubes add 1 mL 0.1 M NaBr. Then to each tube add four drops 0.1 M $AgNO_3$. A very insoluble precipitate of AgBr will form. To the first tube add 0.1 M $Na_2S_2O_3$, drop by drop, until you have added about 1 mL. If the silver thiosulfate complex ion is more stable than AgBr, the complex will form and the solid will dissolve. Repeat the tests with the Na_2SO_3 and Na_2SO_4 solutions and record your results.

5. PROPERTIES OF THE PREPARED SOLID. In Part A of this experiment you prepared a compound that is supposed to be $Na_2S_2O_3 \cdot 5 H_2O$. Weigh the solid, which should be about dry by now, to ± 0.1 g, on its piece of filter paper. Then weigh a piece of filter paper and calculate the amount of solid you obtained. Also weigh the piece of filter paper on which you have the excess sulfur, and find the mass of the unreacted sulfur.

In this experiment you carried out several experiments on solutions containing $S_2O_3^{2-}$, SO_3^{2-}, and SO_4^{2-} ions. Using your observations in these experiments, carry out four tests to establish that your prepared solid contains $S_2O_3^{2-}$ ions. Then determine, by at least one test, whether the solid also contains SO_3^{2-} and/or SO_4^{2-} ion.

DISPOSAL OF REAGENTS. For the most part the reagents used in this experiment are either nonpolluting or used in very small amounts. Dispose of the contents of the waste beaker and the suction flask as directed by your instructor.

Data and Observations: Sulfur Chemistry

A. Preparation of Sodium Thiosulfate

Mass of Na_2SO_3 _____ g   Mass of sulfur   _____ g

Mass of sodium thiosulfate plus filter paper   _____ g

Mass of excess sulfur plus filter paper   _____ g

Mass of filter paper _____ g   Mass of $Na_2S_2O_3 \cdot 5\ H_2O$ _____ g

Mass of excess sulfur _____ g   Mass of sulfur that reacted _____ g

Moles of sulfur that
reacted   _____ moles   Moles of $Na_2S_2O_3 \cdot 5\ H_2O$
produced by reaction   _____ moles

Moles of
$Na_2S_2O_3 \cdot 5\ H_2O$
that were recovered
(MM = 248.2 g)   _____ moles   Percentage yield   _____ %

B. Properties of Sulfur Oxyanions

1. Behavior on addition of 6 M HCl

$S_2O_3{}^{2-}$ $SO_3{}^{2-}$ $SO_4{}^{2-}$

Observations:

Net ionic equations for any reactions that occurred:

2. Behavior on addition of 1 M $BaCl_2$, followed by 6 M HCl

$S_2O_3{}^{2-}$ $SO_3{}^{2-}$ $SO_4{}^{2-}$

Observations:

Net ionic equations for any precipitation reactions that occurred:

Net ionic equations for any reactions between precipitates and 6 M HCl:

(continued on following page)

(continued)

3. Redox behavior on addition of I_2

$S_2O_3^{2-}$ $\qquad\qquad\qquad$ SO_3^{2-} $\qquad\qquad\qquad$ SO_4^{2-}

Observations:

Net ionic equations for any reactions that occurred:

4. Complex ion formation on addition to Ag^+ ion

$S_2O_3^{2-}$ $\qquad\qquad\qquad$ SO_3^{2-} $\qquad\qquad\qquad$ SO_4^{2-}

Observations:

Net ionic equations for any reactions that occurred (coordination no. is 2):

Did any oxyanion tend to reduce Ag^+ ion? _____ Which one, if any? _____
Rank the oxyanions in order of increasing strength as a reducing agent.

_____ _____ _____

Behavior on addition of oxyanions to AgBr

$S_2O_3^{2-}$ $\qquad\qquad\qquad$ SO_3^{2-} $\qquad\qquad\qquad$ SO_4^{2-}

Observations:

Net ionic equations for any reactions that occurred:

Rank the oxyanions in order of increasing stability of silver oxyanion complex.

_____ _____ _____

5. Tests used to confirm that prepared compound contains $S_2O_3^{2-}$ ion

1. $\qquad\qquad\qquad\qquad$ 2.

3. $\qquad\qquad\qquad\qquad$ 4.

Conclusions:

Tests used to test for presence of SO_3^{2-} and/or SO_4^{2-}

1. $\qquad\qquad\qquad\qquad$ 2.

Conclusions:

Advance Study Assignment: Sulfur Chemistry

1. Some of the sulfur-containing species discussed in this experiment are the $S_2O_3^{2-}$, S_5^{2-}, and $S_4O_6^{2-}$ ions. Draw the Lewis structures for these ions.

2. Given the Lewis structures from Problem 1, predict the following:

    a. The S—S—S bond angle in S_5^{2-}.   ——————— °

    b. The S—S—S bond angle in $S_4O_6^{2-}$.   ——————— °

3. Write the balanced equation for the reaction by which we make thiosulfate ion, $S_2O_3^{2-}$, from sulfur, S_8, and sulfite ion, SO_3^{2-}, in this experiment.

4. Given the equation in Problem 3 and information in the discussion section of this experiment, explain why it is reasonable that $S_2O_3^{2-}$ ion is stable in distilled water but will decompose if a strong acid is added. The decomposition occurs by the reverse of the formation reaction.

5. 6.0 g of Na_2SO_3 are dissolved in water and 2.0 g of sulfur are added. After boiling the mixture for 10 minutes, a student finds that 1.0 g of sulfur has reacted with sulfite ion, producing thiosulfate ion. Molar masses in grams: S_8, 256.3; Na_2SO_3, 126.1; $S_2O_3^{2-}$, 112.1; $Na_2S_2O_3 \cdot 5\ H_2O$, 248.2.

    a. How many moles of S_8 reacted?

                                                          ——————— moles

    b. How many moles of $S_2O_3^{2-}$ ion were formed?

                                                          ——————— moles

    c. After the excess sulfur was removed and the solution cooled to 2°C, 3.0 g $Na_2S_2O_3 \cdot 5$ H_2O crystallized out and were recovered. What percent of the thiosulfate ion produced in the reaction was actually obtained as product?

                                                          ——————— %

Preparation of Aspirin

One of the simpler organic reactions that one can carry out is the formation of an ester from an acid and an alcohol:

$$R-\overset{\overset{\displaystyle O}{\|}}{C}-OH + HO-R' \rightarrow R-\overset{\overset{\displaystyle O}{\|}}{C}-O-R' + H_2O \qquad (1)$$

an acid   an alcohol   an ester

In the equation, R and R' are H atoms or organic fragments like CH_3, C_2H_5, or more complex aromatic groups. There are many esters, because there are many organic acids and alcohols, but they all can be formed, in principle at least, by Reaction 1. The driving force for the reaction is, in general, not very great, so that one ends up with an equilibrium mixture of ester, water, and acid, and alcohol.

There are some esters that are solids because of their high molecular weight or other properties. Most of these esters are not soluble in water, so they can be separated from the mixture by crystallization. This experiment deals with an ester of this sort, the substance commonly called aspirin. Aspirin is the active component in headache pills and is one of the most effective, relatively nontoxic, pain killers.

Aspirin can be made by the reaction of the —OH group in the salicyclic acid molecule with the carboxyl (—COOH) group in acetic acid:

acetic acid   salicylic acid   aspirin

A better preparative method, which we will use in this experiment, employs acetic anhydride in the reaction instead of acetic acid. The anhydride can be considered to be the product of a reaction in which two acetic acid molecules combine, with the elimination of a molecule of water. The anhydride will react with the water produced in the esterification reaction and will tend to drive the reaction to the right. A catalyst, normally sulfuric or phosphoric acid, is also used to speed up the reaction.

acetic anhydride   salicylic acid   aspirin   acetic acid

The aspirin you will prepare in this experiment is relatively impure and should certainly not be taken internally, even if the experiment gives you a bad headache.

There are several ways by which the purity of your aspirin can be estimated. Probably the simplest way is to measure its melting point. If the aspirin is pure, it will melt sharply at the literature value of the melting point. If it is impure, the melting point will be lower than the literature value by an amount that is roughly proportional to the amount of impurity present.

A more quantitative measure of the purity of your aspirin sample can be obtained by determining the percentage of salicylic acid it contains. Salicylic acid is the most likely impurity in the sample because, unlike acetic acid, it is not very soluble in water. Salicylic acid forms a highly colored magenta complex with Fe(III). By measuring the absorption of light by a solution containing a known amount of aspirin in excess Fe^{3+} ion, one can easily determine the percentage of salicylic acid present in the aspirin.

Experimental Procedure

WEAR YOUR SAFETY GLASSES WHILE
PERFORMING THIS EXPERIMENT

Weigh a 50-mL Erlenmeyer flask on a triple-beam or top-loading balance and add 2.0 g of salicyclic acid. Measure out 5.0 mL of acetic anhydride in your graduated cylinder, and pour it into the flask in such a way as to wash any crystals of salicylic acid on the walls down to the bottom. Add five drops of 85% phosphoric acid to serve as a catalyst. (**CAUTION:** *Both acetic anhydride and phosphoric acid are reactive chemicals that can give you a bad chemical burn, so use due caution in handling them. If you get any of either on your hands or clothes, wash thoroughly with soap and water.*)

Clamp the flask in place in a beaker of water supported on a wire gauze on a ring stand. Heat the water with a Bunsen burner to about 75°C, stirring the liquid in the flask occasionally with a stirring rod. Maintain this temperature for about 15 minutes, by which time the reaction should be complete. *Cautiously,* add 2 mL of water to the flask to decompose any excess acetic anhydride. There will be some hot acetic acid vapor evolved as a result of the decomposition.

When the liquid has stopped giving off vapors, remove the flask from the water bath and add 20 mL of water. Let the flask cool for a few minutes in air, during which time crystals of aspirin should begin to form. Put the flask in an ice bath to hasten crystallization and increase the yield of product. If crystals are slow to appear, it may be helpful to scratch the inside of the flask with a stirring rod. Leave the flask in the ice bath for at least 5 minutes.

Collect the aspirin by filtering the cold liquid through a Buchner funnel using suction. Turn off the suction and pour about 5 mL of ice-cold distilled water over the crystals; after about 15 seconds turn on the suction to remove the wash liquid along with most of the impurities. Repeat the washing process with another 5-mL sample of ice-cold water. Draw air through the funnel for a few minutes to help dry the crystals and then transfer them to a piece of dry, weighed filter paper. Weigh the sample on the paper to ± 0.1 g.

Test the solubility properties of the aspirin by taking samples of the solid the size of a pea on your spatula and putting them in separate 1-mL samples of each of the following solvents and stirring:

1. toluene, $C_6H_5CH_3$, nonpolar aromatic
2. hexane, C_6H_{14}, nonpolar aliphatic
3. ethyl acetate, $C_2H_5OCOCH_3$, aliphatic ester
4. ethyl alcohol, C_2H_5OH, polar aliphatic, hydrogen bonding
5. acetone, CH_3COCH_3, polar aliphatic, nonhydrogen bonding
6. water, highly polar, hydrogen bonding

To determine the melting point of the aspirin, add a small amount of your prepared sample to a melting point tube (made from 5-mm tubing), as directed by your instructor. Shake the solid down by tapping the tube on the bench top, using enough solid to give you a depth of about 5 mm. Set up the apparatus shown in Figure 41.1. Fasten the melting point tube to the thermometer with a small rubber band, which should be above the surface of the oil. The thermometer bulb and sample should be about 2 cm above the bottom of the tube. Heat the oil bath *gently,* especially after the temperature gets above 100°C. As the melting point is approached, the crystals will begin to soften. Report the melting point as the temperature at which the last crystals disappear.

To analyze your aspirin for its salicylic acid impurity, weigh out 0.10 ± 0.01 g of your sample into a weighed 100-mL beaker. Dissolve the solid in 5 mL 95% ethanol. Add 5 mL 0.025 M $Fe(NO_3)_3$ in 0.5 M HCl and 40 mL distilled water. Make all volume measurements with a graduated cylinder. Stir the solution to mix all reagents.

Rinse out a spectrophotometer tube with a few milliliters of the solution and then fill the tube with that solution. Measure the absorbance of the solution at 525 nm. The absorbance measurement should be made *within 5 minutes* of the time the sample was dissolved in the ethanol, since aspirin will gradually decompose in solution, producing salicylic acid and acetic acid. From the calibration curve or equation provided, calculate the percentage of salicylic acid in the aspirin sample.

The contents of the suction flask may be poured down the drain. The toluene and hexane from the solubility tests should be put in the waste crock.

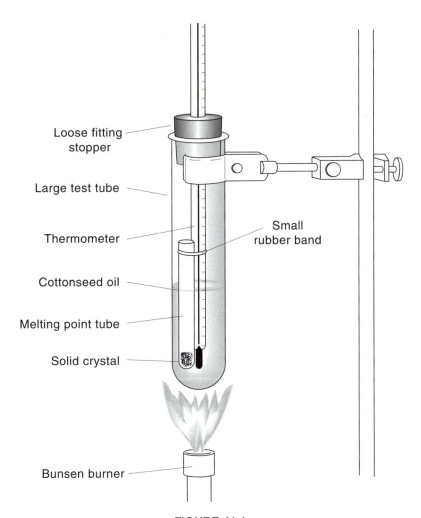

Loose fitting stopper

Large test tube

Thermometer

Small rubber band

Cottonseed oil

Melting point tube

Solid crystal

Bunsen burner

FIGURE 41.1

Data and Results: Preparation of Aspirin

Mass of salicylic acid used   _____ g

Volume of acetic anhydride used   _____ mL

Mass of acetic anhydride used
(density = 1.08 g/mL)   _____ g

Mass of aspirin obtained   _____ g

Theoretical yield of aspirin   _____ g

Percentage yield of aspirin   _____ %

Melting point of aspirin   _____ °C

Absorbance of aspirin solution   _____

Percentage of salicylic acid impurity   _____ %

Solubility properties of aspirin

   Toluene   _____   Ethyl alcohol _____

   Hexane   _____   Acetone   _____

   Ethyl acetate _____   Water   _____

S = soluble   I = insoluble   SS = slightly soluble

Underline those characteristics listed which would be likely to be present in a good solvent
for aspirin.

      organic   aliphatic   polar   hydrogen bonding

      inorganic   aromatic   nonpolar   nonhydrogen bonding

Experiment 41 *Name* ______________________ *Section* __________

Advance Study Assignment: Preparation of Aspirin

1. Calculate the theoretical yield of aspirin to be obtained in this experiment, starting with 2.0 g of salicylic acid and 5.0 mL of acetic anhydride (density = 1.08 g/mL).

______________ g

2. If 1.9 g of aspirin were obtained in this experiment, what would be the percentage yield?

______________ %

3. The name acetic anhydride implies that the compound will react with water to form acetic acid. Write the equation for the reaction.

4. Identify R and R′ in Equation 1 when the ester, aspirin, is made from salicylic acid and acetic acid.

5. Write the equation for the reaction by which aspirin decomposes in an aqueous ethanol solution.

Experiment 42

ANALYSIS FOR VITAMIN C

Vitamin C, known chemically as ascorbic acid, is an important component of a healthy diet. In the mid-eighteenth century the British navy found that the addition of citrus fruit to the sailors' diet prevented the malady called scurvy. Humans are one of the few members of the animal kingdom unable to synthesize vitamin C, resulting in the need for regular ingestion in order to remain healthy. The National Academy of Sciences has established the threshold of 60 mg/day for adults as the Recommended Dietary Allowance (RDA). Linus Pauling, a chemist whose many contributions to chemical bonding theory should be well-known to you, recommended a level of 500 mg/day to help ward off the common cold. He had also suggested that large doses of vitamin C are helpful in preventing cancer.

The vitamin C content of foods can easily be determined by oxidizing ascorbic acid, $C_6H_8O_6$ to dehydro-L-ascorbic acid, $C_6H_6O_6$:

$$C_6H_8O_6 \rightarrow C_6H_6O_6 + 2\,H^+ + 2e^-$$

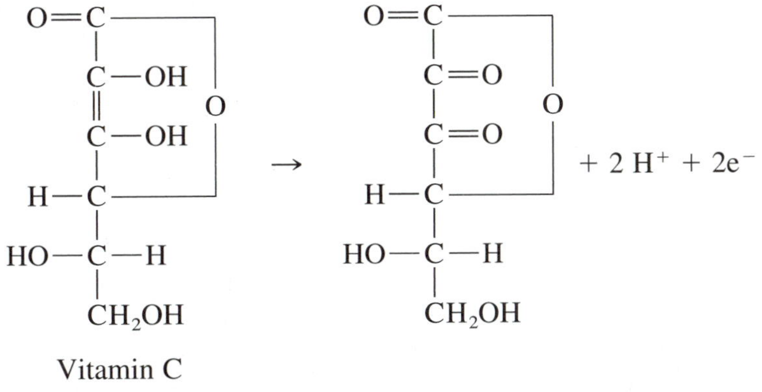

Vitamin C

This reaction is very slow for ascorbic acid in the dry state, but occurs readily when in contact with moisture. A reagent that is particularly good for the oxidation is an aqueous solution of iodine, I_2. Since I_2 is not very soluble in water, we dissolve it in a solution of potassium iodide, KI, in which the I_2 exists mainly as I_3^-, a complex ion. The reaction with ascorbic acid involves I_2, which is reduced to I^- ion.

$$2e^- + I_2 \rightarrow 2\,I^-$$

In the overall reaction, one mole of ascorbic acid requires one mole of I_2 for complete oxidation.

When the red-colored I_2 solution is added to the ascorbic solution, the characteristic iodine color disappears because of the above reaction. Although we could use the first permanent appearance of the yellow color of dilute iodine to mark the end point of the titration, better results are obtained when starch is added as an indicator. Starch reacts with I_2 to form an intensely colored blue complex. In the titration I_2 reacts preferentially with ascorbic acid, and so its concentration remains very low until the ascorbic acid is all oxidized. At that point, the I_2 concentration begins to go up and the reaction with the indicator occurs:

$$I_2 + starch \rightarrow starch\text{-}I_2complex$$

yellow   blue

Because an I_2 solution cannot be prepared accurately by direct weighing, it is necessary to standardize the I_2 against a reference substance of known purity. We will use pure ascorbic acid for this reference, or primary standard. After standardization you can use the iodine solution for the direct determination of vitamin C in any kind of sample.

Experimental Procedure

WEAR YOUR SAFETY GLASSES WHILE
PERFORMING THIS EXPERIMENT

A. Standardization of the Iodine Solution

Obtain from the storeroom a buret and an unknown vitamin C sample. Weigh out accurately on the analytical balance three ascorbic acid samples of approximately 0.10 g into clean 250-mL Erlenmeyer flasks. Dissolve each sample in approximately 100 mL of water.

Clean your buret thoroughly. Draw about 100 mL of the stock I_2 solution from the carboy in the laboratory into a 400-mL beaker and add approximately 150 mL of water. Stir thoroughly and cover with a piece of aluminum foil. Rinse the buret with a few milliliters of the I_2 solution three times. Drain and then fill the buret with the I_2 solution.

After taking an initial reading of the buret (you may find looking toward a light source will make it easier to see the bottom of the meniscus), add 1 mL of starch indicator to the first ascorbic acid sample and titrate with the iodine solution. Note the change of the I_2 color as you swirl the flask gently and continuously during the titration. Continue the addition of the iodine, using progressively smaller volume increments, until the sample solution just turns a distinct blue. After reading the buret, titrate the other two sample solutions—being sure to add the starch indicator and to read your buret before and after each titration.

B. Analysis of an Unknown Containing Vitamin C

Given your experience with the standardization reaction, you should be able to devise an analogous procedure to determine the vitamin C content of your unknown sample. You will need to select a sample size, and you may need to carry out an initial treatment of the sample. In particular, if your instructor assigns you a fruit juice sample, it will be desirable to first filter the sample through cheese cloth, followed by rinsing of the filter with water.

It may be helpful in choosing the sample sizes to calculate an iodine solution parameter called the titer—the number of mg of ascorbic acid which reacts with 1 mL of iodine solution. This number is easily found from the I_2 concentration and the mass relationship in the reaction. It is desirable to have the volume of I_2 for each titration be at least 15 mL. Using a small initial sample will give you an indication of how much to scale up for your final titrations.

Report your results in per cent vitamin C, if a solid sample was used. For liquid samples, report mg of vitamin C per 100 mL. In each case, calculate the size sample required to give the RDA of vitamin C.

Data and Calculations: Analysis for Vitamin C

A. Standardization of Iodine Solution

Sample	*I*	*II*	*III*	
Mass of Ascorbic Acid Sample	_____	_____	_____	g
Moles of Ascorbic Acid	_____	_____	_____	g
Initial buret reading	_____	_____	_____	mL
Final buret reading	_____	_____	_____	mL
Volume of I_2 added	_____	_____	_____	mL
Moles of I_2 consumed	_____	_____	_____	
Molarity of I_2	_____	_____	_____	M
Titer of I_2 (MM Ascorbic Acid = 176 g/mole)	_____	_____	_____	mg/mL

B. Unknown Sample

	I	*II*	*III*	
Mass or volume of unknown	_____	_____	_____	
Initial buret reading	_____	_____	_____	mL
Final buret reading	_____	_____	_____	mL
Moles of iodine added	_____	_____	_____	
Moles of vitamin C in sample	_____	_____	_____	
Mass of vitamin C in sample	_____	_____	_____	mg
Per cent vitamin C in solid sample	_____	_____	_____	
Concentration of vitamin C in liquid sample	_____	_____	_____	
Amount of sample which will furnish RDA	_____	_____	_____	

361

Advance Study Assignment: Analysis for Vitamin C

1. Write a balanced equation for the reaction between I_2 and ascorbic acid. Identify the oxidizing agent and the reducing agent.

2. A solution of I_2 was standardized with ascorbic acid. Using a 0.1000-g sample of pure ascorbic acid, 25.32 mL of I_2 were required to reach the starch end point.

    a. What is the molarity of the iodine solution?

    _____ M

    b. What is the titer of the iodine solution?

    _____ mg asc/mL I_2

3. A sample of fresh grapefruit juice was filtered and titrated with the above I_2 solution. A 100-mL sample of the juice took 9.85 mL of the iodine solution to reach the starch end point.

    a. What is the concentration of vitamin C in the juice in mg vitamin C/100 mL of juice?

    _____ mg/100 mL

    b. What quantity of juice will provide the RDA amount of vitamin C?

    _____ mL

VAPOR PRESSURE OF WATER

Temperature °C	Pressure mm Hg	Temperature °C	Pressure mm Hg
0	4.6	26	25.2
1	4.9	27	26.7
2	5.3	28	28.3
3	5.7	29	30.0
4	6.1	30	31.8
5	6.5	31	33.7
6	7.0	32	35.7
7	7.5	33	37.7
8	8.0	34	39.9
9	8.6	35	42.2
10	9.2	40	55.3
11	9.8	45	71.9
12	10.5	50	92.5
13	11.2	55	118.0
14	12.0	60	149.4
15	12.8	65	187.5
16	13.6	70	233.7
17	14.5	75	289.1
18	15.5	80	355.1
19	16.5	85	433.6
20	17.5	90	525.8
21	18.7	95	633.9
22	19.8	97	682.1
23	21.1	99	733.2
24	22.4	100	760.0
25	23.8	101	787.6

To convert mm Hg to kPa, multiply the entry in the table by 0.1333.

$$1 \text{ mm Hg} = 0.1333 \text{ kPa}$$

SUMMARY OF SOLUBILITY PROPERTIES OF IONS AND SOLIDS

	Cl^-	SO_4^{2-}	CO_3^{2-}, PO_4^{3-}	CrO_4^{2-} yellow	OH^- O^{2-}	H_2S, pH = 0.5	S^{2-}, pH = 9
Na^+, K^+, NH_4^+	S	S	S	S	S	S	S
Ba^{2+}	S	I	A	A	S^-	S	S
Ca^{2+}	S	S^-	A	S	S^-	S	S
Mg^{2+}	S	S	A	S	A	S	S
Fe^{3+} (yellow)	S	S	A	A	A	S	A
Cr^{3+} (blue-violet)	S	S	A	A	A	S	A
Al^{3+}	S	S	A, B	A, B	A, B	S	A, B
Ni^{2+} (green)	S	S	A, N	A, N	A, N	S	A^+, O^+
Co^{2+} (pink)	S	S	A	A	A	S	A^+, O^+
Zn^{2+}	S	S	A, B, N	A, B, N	A, B, N	S	A
Mn^{2+} (lt. pink)	S	S	A	A	A	S	A
Cu^{2+} (blue)	S	S	A, N	A, N	A, N	O	O
Cd^{2+}	S	S	A, N	A, N	A, N	A^+, O	A^+, O
Bi^{3+}	A	A	A	A	A	O	O
Hg^{2+}	S	S	A	A	A	O^+, C	O^+, C
Sn^{2+}, Sn^{4+}	A, B	A, B	A, B	A, B	A, B	A^+, C	A^+, C
Sb^{3+}	A, B	A, B	A, B	A, B	A, B	A^+, C	A^+, C
Ag^+	A^+, N	S^-, N	A, N	A, N	A, N	O	O
Pb^{2+}	HW, B, A^+	B	A, B	B	A, B	O	O
Hg_2^{2+}	O^+	S^-, A	A	A	A	O^+	O^+

Key: S, soluble in water.   I, insoluble in any common reagent.
A, soluble in acid (6 M HCl or other S^-, slightly soluble in water.
nonprecipitating, nonoxidizing acid). A^+, soluble in 12 M HCl.
B, soluble in 6 M NaOH. O^+, soluble in aqua regia.
O, soluble in hot 6 M HNO_3.   C, soluble in 6 M NaOH containing excess S^{2-}.
N, soluble in 6 M NH_3.   HW, soluble in hot water.

Example: For Cu^{2+} and OH^- the entry is A, N. This means that $Cu(OH)_2(s)$, the product obtained when solutions containing Cu^{2+} and OH^- are mixed, will dissolve to the extent of at least 0.1 mole per liter when treated with 6 M HCl or 6 M NH_3. Since 6 M HNO_3, 12 M HCl, and aqua regia are at least as strongly acidic as 6 M HCl, $Cu(OH)_2(s)$ would also be soluble in those reagents.

TABLE OF ATOMIC MASSES
(BASED ON CARBON-12)

	Symbol	Atomic No.	Atomic Mass		Symbol	Atomic No.	Atomic Mass
Actinium	Ac	89	[227]*	Iridium	Ir	77	192.2
Aluminum	Al	13	26.9815	Iron	Fe	26	55.847
Americium	Am	95	[243]	Krypton	Kr	36	83.80
Antimony	Sb	51	121.75	Lanthanum	La	57	138.91
Argon	Ar	18	39.948	Lawrencium	Lw	103	[257]
Arsenic	As	33	74.9216	Lead	Pb	82	207.19
Astatine	At	85	[210]	Lithium	Li	3	6.939
Barium	Ba	56	137.34	Lutetium	Lu	71	174.97
Berkelium	Bk	97	[247]	Magnesium	Mg	12	24.312
Beryllium	Be	4	9.0122	Manganese	Mn	25	54.9380
Bismuth	Bi	83	208.980	Mendelevium	Md	101	[256]
Boron	B	5	10.811	Mercury	Hg	80	200.59
Bromine	Br	35	79.909	Molybdenum	Mo	42	95.94
Cadmium	Cd	48	112.40	Neodymium	Nd	60	144.24
Calcium	Ca	20	40.08	Neon	Ne	10	20.183
Californium	Cf	98	[249]	Neptunium	Np	93	[237]
Carbon	C	6	12.01115	Nickel	Ni	28	58.71
Cerium	Ce	58	140.12	Niobium	Nb	41	92.906
Cesium	Cs	55	132.905	Nitrogen	N	7	14.0067
Chlorine	Cl	17	35.453	Nobelium	No	102	[253]
Chromium	Cr	24	51.996	Osmium	Os	76	190.2
Cobalt	Co	27	58.9332	Oxygen	O	8	15.9994
Copper	Cu	29	63.546	Palladium	Pd	46	106.4
Curium	Cm	96	[247]	Phosphorus	P	15	30.9738
Dysprosium	Dy	66	162.50	Platinum	Pt	78	195.09
Einsteinium	Es	99	[254]	Plutonium	Pu	94	[242]
Erbium	Er	68	167.26	Polonium	Po	84	[210]
Europium	Eu	63	151.96	Potassium	K	19	39.102
Fermium	Fm	100	[253]	Praseodymium	Pr	59	140.907
Fluorine	F	9	18.9984	Promethium	Pm	61	[145]
Francium	Fr	87	[223]	Protactinium	Pa	91	[231]
Gadolinium	Gd	64	157.25	Radium	Ra	88	[226]
Gallium	Ga	31	69.72	Radon	Rn	86	[222]
Germanium	Ge	32	72.59	Rhenium	Re	75	186.2
Gold	Au	79	196.967	Rhodium	Rh	45	102.905
Hafnium	Hf	72	178.49	Rubidium	Rb	37	85.47
Helium	He	2	4.0026	Ruthenium	Ru	44	101.07
Holmium	Ho	67	164.930	Samarium	Sm	62	150.35
Hydrogen	H	1	1.00797	Scandium	Sc	21	44.956
Indium	In	49	114.82	Selenium	Se	34	78.96
Iodine	I	53	126.9044	Silicon	Si	14	28.086

	Symbol	Atomic No.	Atomic Mass		Symbol	Atomic No.	Atomic Mass
Silver	Ag	47	107.870	Tin	Sn	50	118.69
Sodium	Na	11	22.9898	Titanium	Ti	22	47.90
Strontium	Sr	38	87.62	Tungsten	W	74	183.85
Sulfur	S	16	32.064	Uranium	U	92	238.03
Tantalum	Ta	73	180.948	Vanadium	V	23	50.942
Technetium	Tc	43	[99]	Xenon	Xe	54	131.30
Tellurium	Te	52	127.60	Ytterbium	Yb	70	173.04
Terbium	Tb	65	158.924	Yttrium	Y	39	88.905
Thallium	Tl	81	204.37	Zinc	Zn	30	65.37
Thorium	Th	90	232.038	Zirconium	Zr	40	91.22
Thulium	Tm	69	168.934				

A value given in brackets denotes the mass number of the longest-lived or best-known isotope.

Appendix IV

MAKING MEASUREMENTS— LABORATORY TECHNIQUES

A science like chemistry requires that one be able to accurately measure many physical quantities. Among these quantities are mass, volume, temperature, pressure, light absorption, and pH. In the course of the laboratory you will make all of these measurements one or more times. In this appendix we will describe the kinds of apparatus that are used and how to operate them properly.

MASS

One of the most important measurements you will be making involves determination of mass. This is always done by measuring by one means or another the force exerted on the sample by gravity. This force is proportional to the mass of the sample. The device that is used is called a balance, because the mass of the sample is established by balancing it against a standard mass or force. In Figure IV.1 are shown several balances that have been used in this century. The balance at the top was used until about 1950. Using it required a set of weights that were put on one pan until the two pans came to the same height; with care one could weigh a sample to 0.0001 grams, but it would take several minutes to make one weighing. There were a lot of drawbacks to such a balance, but they were used for over a century in the form shown or more primitive ones. After World War II, instrument design developed rapidly. The balance at the left is a mechanical one, but the weights are internal, and added to or taken from a pan inside the balance. Final readings are taken from a set of dials that add or take off weights, and an optically projected scale that furnishes masses in the milligram range. This type of balance is still used in many schools, and if you have one in your laboratory, your instructor will show you how it operates. The balance on the right is called a top-loading balance; mass is read directly from dials and an optical scale; this kind of balance gives rapid results but is limited in its precision to milligrams at best.

In recent years electronic balances have been developed that furnish the mass directly with a digital readout (see Figure IV.2). They do not have any weights, but depend on balancing the sample mass by a magnetic force that can be accurately related to mass. These balances can be accurate to 0.0001 g, and may arrive at the mass of a sample within 30 seconds or less. In a very real sense, they are the ultimate weighing machines. In using such a balance, you first depress the control bar. This will zero the balance, and it will read 0.0000 g. Place your sample on the balance pan, close the balance door, and read the mass when the balance gives a steady reading. If your sample is to be in a container, you can find its mass by weighing the empty container, then depressing the control bar to rezero, or tare, the balance. Then when you add your sample, its mass will appear directly as the digital readout. There are many brands of these balances, so if your lab has one, your instructor will describe the details of the operation of your balance before you use it.

Here are some guidelines for successful balance operation, which apply to the weighing of a sample on any balance:

a

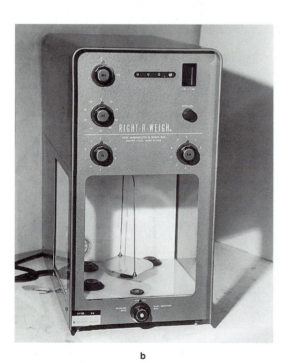

b

c

FIGURE IV.1 **(a)** An old mechanical analytical balance circa 1940—good to 0.0001 g, maybe. **(b)** A modern mechanical analytical balance circa 1960—maximum load 120 g, good to 0.0001 g. **(c)** A modern top-loading semi-automatic balance circa 1970—good to about 1 mg.

1. Be certain the balance has been "zeroed" (reads 0 grams) before you place anything on the balance pan.
2. Never weigh chemicals directly on a balance pan. Always use a suitable container or weighing paper. In several experiments you will weigh a sample tube and measure the mass of sample removed by difference.

FIGURE IV.2 A modern electronic automatic analytical balance, circa 1985. Its maximum load is 120 g, and it is good to 0.0001 g. The beaker weighs 27.5056 g.

3. Be certain that air currents are not disturbing the balance pan. In the case of an analytical balance, always shut the balance case doors when making measurements you are recording.

4. Never put hot, or even warm, objects on the balance pan. The temperature difference will change the density of the air surrounding the balance, and thus give inaccurate measurements.

5. Record your measurement, to the proper number of significant figures, directly on to your Data page or in your notebook.

6. After finishing your measurements, be certain that the balance registers zero again and close the balance doors. Brush out any chemicals which may have been spilled.

7. Be gentle with your balance. It is a sensitive, rather delicate, instrument, and, like a person, responds best when treated properly.

VOLUME

Another very common measurement we make in the laboratory is that of volume, almost always the volume of a liquid. Such measurements are made with varying degrees of accuracy, depending on the situation.

The unit of volume we use is usually the milliliter, 0.001 liters, abbreviated mL, since most volumes we need are between about 1 and 1000 mL. The mL has the same volume as 1 cm^3 or 1 cc, one cubic centimeter.

If we are asked to add 150 mL of water to a beaker during an experiment in which we are carrying out a chemical reaction, we don't need high accuracy, and might use a beaker or flask on which there are some volume markings (as shown in Figure IV.3), which are good to about 5%. Somewhat more precise volumes are obtained with a graduated cylinder, where one can get within about 1% of a needed volume.

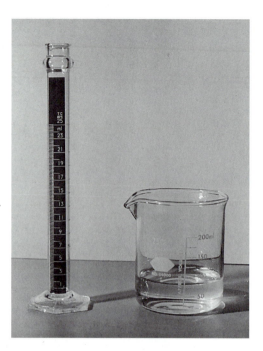

FIGURE IV.3 A 25-mL graduated cylinder, good to 0.2 mL, and a 250-mL beaker with graduations, good to 20 mL.

A. Pipets

More precise volume measurements are often required; these may be obtained with pipets, which are available in many different volumes, from 1 mL up to about 100 mL. In several experiments in this manual pipets are used. Pipets are calibrated to deliver the specified volume when the meniscus at the liquid level is coincident with the horizontal line etched around the upper pipet stem. The following steps make for proper use of a pipet:

1. The pipet must be clean. When it drains, there should be no drops left on the pipet wall. It does not need to be dry.
2. Always use a pipet bulb, not your mouth, when pulling liquid into a pipet. You shouldn't run the risk of getting the liquid into your mouth.
3. Place the pipet bulb on the upper end of the pipet and squeeze the air out. Immerse the tip of the pipet into the liquid and draw up enough liquid to get some into the main body of the pipet. Remove the bulb and place your finger over the top of the pipet stem. Hold the pipet in a horizontal position and swirl the liquid around inside, rinsing the upper stem and body. Drain the liquid into a beaker; repeat the rinsing twice with small volumes of the liquid. This ensures that the contents of the pipet will be the liquid you wish to work with, and not water or a previously-used reagent. Finally, using the bulb, fill the pipet, drawing liquid a centimeter or two above the calibration mark. Put your finger over the top of the stem and, carefully, let liquid flow out into a beaker until the bottom of the meniscus is just at the level of the calibration mark. Wipe off the lower end of the pipet with a tissue and place the tip inside the receiving container, with the tip touching the wall. Release the liquid and let the liquid flow into the container. Hold the tip against the wall for about 30 seconds after it appears that all the liquid has been transferred, to make sure all the liquid drains out.

It takes skill and practice to make good use of a pipet. It is now possible to obtain pipettors, which automatically and repeatedly deliver precise volumes of reagents. These speed up the pipetting procedure enormously, and you may have them in your laboratory.

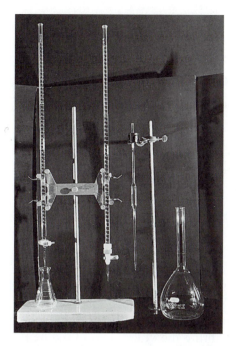

FIGURE IV.4 A pair of 50-mL burets, a 10-mL pipet, and a 1000-mL volumetric flask. The burets and pipet are good to 0.01 mL, and the flask is good to about 0.1 mL.

B. Burets

A buret is a long calibrated tube fitted with a stopcock to control release of reagent liquids. Burets are used in a procedure called a titration, and are frequently used in pairs. This procedure allows one to add precisely measured volumes of reagents to a so-called "end point," at which the amount of one reagent is equivalent chemically to another, which may also be a liquid in a buret, or a weighed solid. We have several titration experiments in this manual.

As with a pipet, working with a buret requires skill and practice. The following procedure should be helpful:

1. Check the buret for cleanliness by filling it with distilled water and allowing it to drain out. A clean buret will leave an unbroken film of water on the interior walls, with no greasy beads. If necessary, clean your buret with soapy water and a long buret brush. Rinse with tap water, and finally with distilled water.

2. Rinse the buret three times with a few milliliters of the reagent to be used. Tip the buret to wash the walls with the reagent, and drain through the tip into a beaker. With the stopcock closed, fill the buret with reagent to a level a little above the top graduation, making sure to fill the tip completely. Open the stopcock carefully and let the liquid level go just below the top zero line. Read the level to the nearest 0.01 mL. To do this, place a white card with a sharp black rectangle on it (a piece of black tape is ideal) behind the buret, so that the reflection of the bottom of the meniscus is just above the upper black line, and at eye level (see Figure IV.5). Then make the volume reading.

   Add the reagent until you get to the end point of the titration, which is usually established by a color change of a chemical called an indicator. To hit the end point within one drop, you need to add reagent slowly when the titration is nearly complete, more rapidly at the beginning. Swirl the container to ensure the reactants are well-mixed. Often you get a clue from the indicator that you are near the end point. If you go past it, you can use the first titration as a guide for the second or third trial. Read the liquid level as before to 0.01 mL.

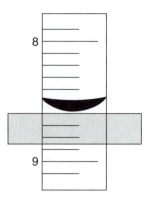

FIGURE IV.5 Reading a buret. The volume reading is 8.56 mL.

C. Volumetric Flasks

A volumetric flask is one which has been especially calibrated to hold a specified volume: 10.00 mL, 25.00 mL, etc. These flasks are used when accurate dilutions are required in analytical experiments. Place the solute or solution in a previously cleaned, but not necessarily dry, flask. Add water to bring the volume up to the mark on the flask. After the liquid volume reaches the lower neck of the flask, add the water with a wash bottle. The last volume increments should be added dropwise until the bottom of the meniscus is coincident with the mark on the flask. Stopper the flask and mix thoroughly by inverting 20 times.

D. Pycnometers

The most precise volume measurements are made with a pycnometer, such as the one used in the first experiment. A pycnometer is simply a container with a very well-defined volume, such as a small flask fitted with a ground glass stopper. With such a device, and a liquid with an accurately known density, one can determine the volume of the pycnometer to 0.001 mL, which is much better than you can do with a pipet or buret. Knowing that volume, you can measure the density of an unknown liquid with an accuracy equal to that of the density of the calibrating liquid, usually within 0.01%.

TEMPERATURE

Many chemical experiments require a knowledge of, or control of, the temperature of the sample under study. This is always done with a thermometer. Your lab drawer probably contains a standard laboratory thermometer suitable for temperature measurements between about $-10°C$ and $150°C$, with $1°$ graduations. We will use this thermometer for most of the experiments in which temperature measurement is needed.

The typical thermometer contains a bulb connected to a fine capillary tube. It contains a liquid, usually mercury, which fills the bulb and part of the capillary. As the temperature goes up, the mercury expands and rises in the capillary. To make the graduations, one would note the level at $0°C$ and $100°C$ and divide that interval into 100 equal parts. The scale fits the one obtained from ideal gas behavior quite well. Your standard thermometer can be read to about $0.2°C$ if you are careful, but given the way it is made, the actual error in the temperature is likely to be greater than that. Other liquids are sometimes used in thermometers (alcohol is common), but they typically do not furnish as reliable temperatures as Hg thermometers.

In some experiments it is desirable to be able to get more precise temperature values than are obtainable from the standard thermometer. You may be furnished with a thermometer with a more limited range, which will read directly to $0.1°C$.

Thermometers are fragile and relatively expensive, so be careful when using them. If you slip a split rubber stopper around the thermometer and support it with a clamp you will minimize breakage, and lab fees. If you break a mercury thermometer, contact your lab supervisor. Liquid mercury has a low vapor pressure, but that vapor is very toxic, so it is important to clean up mercury spills completely, which is not easy. With limited-range thermometers, do not heat them above their maximum readable temperature, since that can ultimately break the thermometer.

When making temperature readings, allow enough time for the level to become steady before noting the final temperature. This will probably take less than a minute, so if you check the reading over a period of time you should be able to get a reliable value.

To obtain temperatures with an accuracy better than 0.1°C is not a simple matter. Some mercury thermometers read to 0.01°C, but they have large bulbs and very fine capillaries and are very expensive, about $250 each. Still higher precision can be obtained using thermistors, which have an electrical resistance that changes rapidly with temperature. By making careful resistance measurements, one can get temperatures accurate to about 0.001°C.

PRESSURE

In studying the behavior of gases or vapors one must be able to measure the gas pressure. In the chemistry lab this is usually done by comparing the pressure with that exerted by the atmosphere, using a device called a manometer. A manometer is a glass U-tube partially filled with a liquid, usually mercury.

A mercury barometer is made from a glass tube which is initially completely filled with mercury. The tube is then inverted while the open end is immersed in a pool of mercury. The level in the tube will fall until the pressure it exerts at the mercury level in the pool is equal to the atmospheric, or barometric, pressure. At that level the pressures inside and outside the glass tube must be equal. Outside, the pressure is that of the atmosphere; inside, it is the pressure exerted by the mercury column, since above the column there is essentially zero pressure.

In a mercury barometer, the glass tube is enclosed in a metal jacket on which there is a scale showing the height in millimeters above the mercury pool. The barometric pressure is equal to the height of the Hg column inside the tube. One can also measure barometric pressure with an aneroid barometer, which consists of a spirally-wound, flexible-walled flat tube containing a fixed amount of air. As the air pressure changes, the tube flexes, which turns a needle that records the pressure.

Barometric pressure is often reported in mm Hg, but it may be given in atmospheres by comparing the value in mm Hg to 760 mm Hg, the standard atmospheric pressure. In SI the pressure is given in either Pascals or bars. One standard atmosphere equals 0.01325×10^{-5} Pascals, or 1.01325 bars. In this book we will usually use pressures in mm Hg, since those are most easily measured directly. In Experiment 7 you may determine the barometric pressure in mm H_2O and then convert it to mm Hg, using the conversion factor: 1 mm Hg $\simeq$ 13.57 mm H_2O.

Having found the barometric pressure, it is a simple matter to measure the gas pressure in a flask such as that shown in Figure IV.6. The pressure inside the flask is equal to the barometric pressure plus the pressure exerted by a mercury column of length $h_2 - h_1$, the Hg heights in the left and right arms of the U-tube. If you know those heights in mm,

$$P_{gas} = (P_{bar} + h_2 - h_1) \text{ mm Hg}$$

When reading the Hg levels in a manometer, take the height at the top of the mercury meniscus in the two arms, which you can read to 1 mm or better.

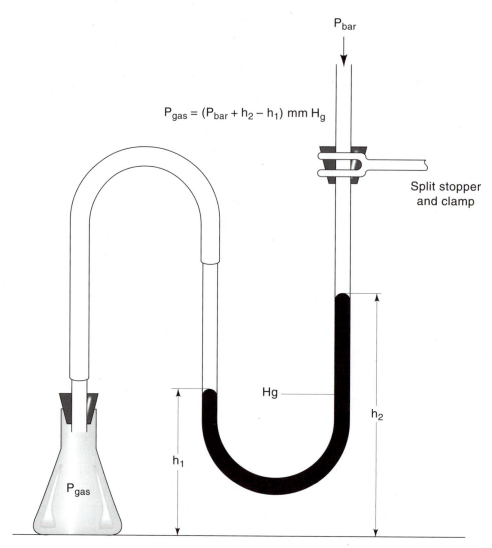

P_{bar}

$P_{gas} = (P_{bar} + h_2 - h_1)$ mm H_g

Split stopper
and clamp

Hg

h_2

h_1

P_{gas}

FIGURE IV.6 Measuring the pressure of a gas.

The U-tube manometer must be handled with care. The long arm should be put into a split stopper and held securely with a clamp. Keep a plastic bucket below the manometer. A mercury spill is a serious matter, since mercury vapor is bad for your health. So be careful and you should have no problems. Plug the ends of the manometer with small corks when you are finished using it.

LIGHT ABSORPTION

In many chemical experiments the interaction of light with a sample can be used to furnish very useful information. In this laboratory course we will often determine the concentration of a species in a solution by measuring the amount of light at a given wavelength (color) that is absorbed by the sample. In an instrument called a spectrophotometer one can do this very easily. The spectrophotometer contains a device called a monochromator (usually made from a diffraction grating), which allows only one wavelength of light to pass through a sample. The absorbance, A, of the sample at that wavelength can be read directly from a meter on the instrument. By a famous equation called Beer's Law, the absorbance is proportional to the concentration, c, of the sample, usually the molarity:

$$A = K \times c$$

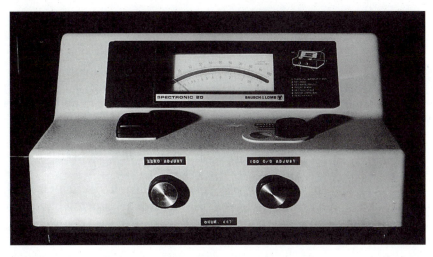

FIGURE IV.7 A Spectronic 20 spectrophotometer. The absorbance of a solution can be read to about 1%.

where K is a constant that depends on the sample, the container, and the wavelength of the light. Most samples obey Beer's Law. To find the concentration of a species in a sample, one measures the absorbance at an appropriate wavelength of a solution containing a known concentration of the species. From that absorbance, or several absorbances obtained with other concentrations, we can make a calibrating graph of concentration vs. absorbance and then use that graph to analyze unknown solutions containing that species. Your instructor will probably furnish you with such a graph in some of the experiments you perform.

A commonly used spectrophotometer is a Spectronic 20 (see Figure IV.7). This instrument has two knobs on the front face. On the left is a zero transmittance adjustment knob and on the right a 100% transmittance adjustment knob. On the top surface at the left is a covered sample chamber, and at the right there is a wavelength selection knob. The Absorbance is read from the upper meter or display register.

To use the spectrophotometer, turn it on about 15 minutes before you need to operate it, to give the components time to warm up. Note the wavelength setting, which is one where the sample absorbs a moderate amount of the incident light. Since the sample is probably colored, the wavelength will most likely be in the visible region of the light spectrum. With the sample compartment closed, turn the zero adjustment knob until the needle or display register indicates 0% transmission. Open the sample chamber and insert a sample tube that is about 3/4 full of a reference solution (usually pure water). Shut the sample chamber and turn the 100% adjustment knob until the needle reads 100% transmission (this will be zero on the absorbance scale). Repeat the adjustment steps until no further changes are necessary. Fill the sample tube about 3/4 full of the solution being studied, insert it in the sample chamber, and close the chamber. Read the Absorbance from the meter or register. Then use the calibration graph to obtain the concentration of the species in the solution.

pH MEASUREMENTS

The pH of a solution is used to describe the concentration of the H^+ ion in that solution. To determine pH one can use acid-base indicators, which change color as the pH changes. Each indicator has a characteristic pH at which the color change occurs, so can be used to indicate whether the pH is higher, or lower, than the characteristic value. By employing several indicators one can fix the pH of a solution within a few tenths of a unit.

Most accurate pH measurements are made with a pH meter. The pH meter is a very high resistance voltmeter which measures the difference in voltage between a reference electrode (usually a so-called calomel electrode) and a glass indicating electrode whose potential is a function of the H^+ ion concentration in the solution in which it is immersed. The meter indicates the pH directly. The two electrodes may be separate, but more commonly are together in one combination electrode. The electrode is kept wet in an appropriate storage solution when not in use.

When you are asked to find the pH of a solution, you may first need to calibrate the pH meter, although it is likely that that was done by your lab supervisor prior to the laboratory session. To calibrate the meter you will need a buffer solution with a well-defined pH as a reference. Put about 25 mL of the buffer in a small beaker. Remove the electrode from the storage solution, rinse it with a stream of distilled water from your wash bottle, and then blot the electrode with a clean tissue. Place the electrode in the buffer. (The electrode is fragile, and expensive, so treat it gently.) Allow the system to equilibrate for a minute or two, or until the pH reading becomes steady. Then use the pH adjust knob to bring the displayed pH to that of the reference buffer solution.

Remove the electrode from the buffer, rinse it with distilled water as before, blot with a tissue, and place the electrode in the solution being studied. Allow time for equilibration, and record the pH. Rinse the electrode and return it to the storage solution.

SEPARATION OF PRECIPITATES FROM SOLUTION

A. Decantation

Decantation is used to remove a liquid from a precipitate by pouring it off. Allow the solution to set for a period of time until the precipitate is on the bottom of the beaker. Position your stirring rod across the beaker with one end protruding beyond the lip. With the index finger of one hand holding the rod in place, pour the liquid slowly down the stirring rod into a receiving vessel. Try not to disturb the precipitate as the last bits of liquid are poured off.

B. Filtration—The Buchner Funnel

Filtration is used when you wish to recover the precipitate in pure form. This is required in Experiment 3. You will use a Buchner funnel, which contains a piece of filter paper of appropriate size covering the holes at the base of the funnel. The Buchner funnel is connected by a rubber stopper or adaptor to the top of a side-arm filter flask. The side arm of the flask is connected via a series of rubber tubes and a safety trap to a water aspirator. When the water is turned on, a vacuum is created in the system by the water rushing by the side opening in the aspirator (see Figure 3.2).

When you are ready to filter, moisten the filter paper with water, then turn the faucet on to start the vacuum. With the water faucet in the fully open position, pour the slurry of solution and precipitate down a stirring rod, as described under decantation. Wash out any remaining precipitate with a slow stream of liquid from your wash bottle. The procedure may call for washing the precipitate on the Buchner funnel with an appropriate volatile liquid before passing air through the filter cake for several minutes to complete drying. Turn off the water, disconnect the tube from the filter flasks, and remove the funnel.

C. Centrifugation

Centrifugation is used to aid in separation of a precipitate from a solution in a test tube. Using the centrifugal force generated by spinning the test tube at several hundred revolutions per minute gives an outward force for settling that is much higher than the gravitational force (see Figure IV.8).

FIGURE IV.8 A lab centrifuge.

To use the centrifuge, be certain that the test tube containing the precipitate is not overly full—the liquid should be at least 3 cm below the top of the tube. Check to see that the test tube has no cracks, as these will cause the tube to break during centrifugation. Place the test tube containing your precipitate and solution in one of the centrifuge tubes. Place a blank tube containing the same amount of water as you have in your sample tube in the centrifuge tube that is opposite your sample tube. Turn on the centrifuge, allow it to spin for about a minute, and then turn it off. Keep your hands away from the spinning centrifuge top. After the spinning top has come to rest, remove the tube. Decant the supernatant liquid from the precipitate.

QUALITATIVE ANALYSIS—A FEW SUGGESTIONS

In the course of your laboratory program you may do several experiments involving qualitative analysis. In those experiments we will use small test tubes and small beakers as sample containers. Separations of precipitates from solutions are always done by centrifugation.

In many procedures you will be told to add 1 mL, or 0.5 mL, to a mixture. This is best accomplished by first finding out what 1 mL volume looks like in a test tube. So, measure out 1 mL in a small graduated cylinder, and pour that into the test tube, noting the height reached by the liquid in the tube. From then on, use that height to tell you the volume to add when you need 1 mL. Half that height indicates 0.5 mL.

Many of the steps in qualitative analysis involve heating a mixture in a boiling-water bath. To make the bath, fill a 250-mL beaker about 2/3 full of water. Support the beaker on a wire screen on an iron ring. Heat the water to the boiling point with your Bunsen burner, and then adjust the flame so as to keep the water at the simmering point. Use this bath to heat the test tube when that is called for. You can remove the tube with your test tube holder.

In separating a solid from a solution, we first centrifuge the mixture. The supernatant liquid is decanted into a test tube or discarded, depending on the procedure. The solid must then be washed free of any remaining liquid. To do this add the indicated wash liquid and stir with a glass stirring rod. Centrifuge again and discard the wash. Keep a set of stirring rods in a 400-mL beaker filled with distilled water. Return a rod to the beaker when you are done using it, and it will be ready when you next need it.

Pay attention to what you are doing, and don't just follow the directions as though you were making a cake. Try to keep in mind what happens to the various cations in each step, so that you won't end up pouring the material you want down the drain and keeping the trash. When you need to put a fraction aside while you are working on another part of the mixture, make sure you know where you put it by labeling the test tube holder with the number of the step in which you will return to analyze that fraction.

Appendix V

MATHEMATICAL CONSIDERATIONS— MAKING GRAPHS

In the laboratory you will be carrying out experiments of various sorts. Some of them will involve almost no mathematics. In others, you will need to make calculations based on your experimental results. The calculations are not difficult, but it is important you make them properly. In particular, you should not report a result that implies an accuracy that is greater, or smaller, than is consistent with the accuracy of your experimental data. In making such calculations we resort to the use of significant figures as a guide to proper reporting of our results. In the first part of this appendix we will discuss significant figures and how to use them.

In several experiments we will carry out a series of measurements on the same system under different conditions. For example, in Experiment 7 we measure the pressure of a sample of gas at different temperatures. In such experiments it is helpful to draw a graph representing our data, since it may reveal some properties of the sample that are not at all obvious if we just have a table of data. The second part of this appendix will show how to interpret a graph and how to construct one properly.

SIGNIFICANT FIGURES

In the laboratory we make many kinds of measurements. The precision of the measurement depends on the device we use to make it. Using an analytical balance, we can measure mass to ± 0.0001 g, so if we have a sample weighing 2.4965 g, we have *five* meaningful figures in our result. These figures are called, reasonably enough, significant figures. If we measure the volume of a sample using a graduated cylinder, the volume we obtain depends on the cylinder we use. If we have a volume of about 6 mL and measure it with a 100-mL cylinder, we would have difficulty distinguishing between a volume of 6 mL and 7 mL, and would be unable to say more than that the volume was about 6 mL. That volume contains only *one* significant figure. With a 10-mL cylinder, we could measure the volume more precisely, report a volume of 6.4 mL, and have confidence that both figures in our result were meaningful, and that it contained *two* significant figures. Many measurements, perhaps most, can be interpreted as we have here.

Sometimes it is not clear how many significant figures there are in a number. Say, for example, we are told to add 1 mL of liquid from a pipet to a test tube. A pipet is a precisely made device, and can deliver 1.00 mL of liquid if used properly. In such a case, it would be sensible to say that there are really three significant figures in that volume, rather than one. It would probably have been better to be told to add 1.00 mL with a pipet, but often that is just not done, even by the authors of this manual.

Some numbers are exact and have no inherent error. There will be an integral number, like 12, or 19, students in your lab group. There is no way there will be 14.5 students, unless there is something strange going on. Conversion factors, used to convert one set of units to another, often contain exact numbers: 1 meter = 100 cm. Both numbers are exact, with no error at all. With volumes, 1 liter = 1000-mL. Again, both numbers are exact. If we convert from one system to another, then usually only one of the numbers in the conversion factor is exact:

1 mole $= 6.022 \times 10^{23}$ molecules. Here the 1 is exact, but the second number is not, and has four significant figures.

If you are in doubt as to the number of significant figures in a given number, there is a method for finding out that usually works. We write the number in exponential notation, expressing the number as the product of a number between 1 and 10 times a power of ten.

$$26.042 = 2.6042 \times 10^1 \quad 0.0091 = 9.1 \times 10^3 \quad 605.20 = 6.0520 \times 10^2$$

The first number in the product has the number of significant figures we seek. So, the first of the above numbers has 5 significant figures, the second has 2, and the third has 5 (a trailing zero is significant). There is one problem with this approach, and that is that sometimes we are given a number with no decimal point, like 2400. We cannot be sure how many significant figures are in that number, since, if the number were rough, there might be 2, but there could be 4 or even more. Here, you have to use some judgment, but in the absence of any guide, you would choose 4.

Students usually have little trouble deciding on the number of significant figures in a piece of data. The difficulty arises when the piece of data is used in a calculation, and they have to express the result using the proper number of significant figures. Say, for example, you are weighing a sample in a beaker and you find that:

> Mass of sample plus beaker = 25.4329 g
>
> Mass of beaker   = 24.6263 g
>
> Mass of sample   = 0.7066 g

Even though the two measured masses contain 6 significant figures, the mass of the sample contains only 4. In another case, where we mix some components to make a solution, weighing each of the components separately on different balances, we get:

> Mass of beaker   = 25.5329 g
>
> Mass of water   = 14.0   g
>
> Mass of salt   = 6.42   g
>
> Total mass   =   ?

Here we have several pieces of data, each with a different degree of precision. The first mass is good to 0.0001 g, the second to 0.1 g, and the third to 0.01 g. If we take the sum, we get 46.9529 g, but we certainly can't report that value, since it could be off, not by 0.0001 g, but by 0.1 g, the possible error in the mass of the water. We can't improve the quality of a result by a calculation, so, using good sense, we can only report the mass to ± 0.1 g. Since the measured mass is closer to 47.0 g than to 46.9 g, we round up to 47.0 as the reported mass, and have a mass with 3 significant figures. Generalizing, *in adding or subtracting numbers, round off the result so that it has the same number of decimal places as there are in the measured quantity with the smallest number of decimal places.* Round the last digit up if the number that follows is greater than 5, and don't round up if it is not.

When *multiplying* or *dividing* measured quantities, the rule is quite simple: *The number of significant figures in the result is equal to the number of significant figures in the quantity with the smallest number of significant figures.*

In the first experiment we measure the density of a liquid using a pycnometer, which is simply a flask with a well-defined volume. We find that volume by weighing the empty pycnometer and then weighing the pycnometer when it is full of water. We might obtain the following data:

Mass of pycnometer plus water   60.8867 g

Mass of empty pycnometer   31.9342 g

Mass of water   28.9525 g

We are asked to find the volume of the pycnometer, given that the density of water is 0.9973 g/mL.

$$density = \frac{mass}{volume} \quad so \quad volume = \frac{mass}{density} = \frac{28.9525 \text{ g}}{0.9973 \text{ g/mL}} = 29.030883 \text{ mL}$$

The mass contains 6 significant figures, the density has 4, and the results from a calculator have 8 or more. Since the density has the smallest number of significant figures, four, we report the volume to 4 significant figures, as 29.03 mL.

In most cases, the reasoning that is required to properly report a calculated value of a property is no more complicated than that we used here. Remember, you can't improve the accuracy in a result by means of a calculation. If you note the quality of your data, you should be able to report your result with the same quality. As with most activities, a little practice helps, and so does a little good sense.

GRAPHING TECHNIQUES

In many chemical experiments we find that one of our measured quantities is dependent on another. If we change one, we change the other. In such an experiment we find data under several sets of conditions, with each data point associated with two measured quantities. In such cases, we can represent these points on a two-dimensional graph, which shows in a continuous way how one quantity depends on another.

Interpreting a Graph

The first graph in the lab manual is in Experiment 3, and is a graph of this kind. It is shown in Figure V.1.

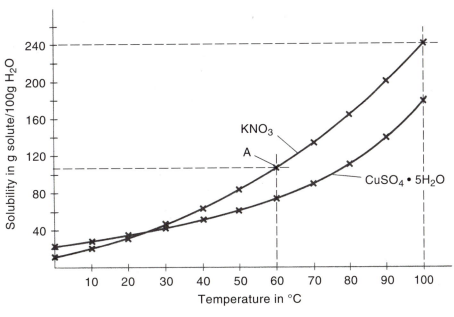

FIGURE V.1

The graph describes how the solubility of each of two compounds, KNO_3 and $CuSO_4 \cdot 5\,H_2O$ depend on temperature. The temperature is shown along the horizontal axis, called the abscissa, over the range from 0 to 100°C. The solubility, in grams per 100 grams water, is along the vertical axis, sometimes called the ordinate. The x's along each of the two curved lines in the graph are data points, showing measured solubility at given temperatures.

The graph has some implied features that are not shown. We actually made the graph on a piece of graph paper on which there was a grid of parallel lines. On the vertical lines the temperature is constant. If you draw a vertical line upwards from the 50°C hashmark on the vertical axis, the temperature on that line is always 50°C. On the vertical line at 73°C, three tenths of the distance between the 70 and 80 hashmarks, it is always 73°C, and so on. Along horizontal lines the solubility is constant. The horizontal line drawn through the 80 hashmark represents a solubility of 80 grams in 100 grams of water all along its length. When we obtained the data for the solubility of KNO_3, we found that, at 60°C, we could dissolve 105 g KNO_3 in 100 g of water. We entered the data point we show as A on the graph at 60°C and 105 g KNO_3. The rest of the data points for the two compounds were entered in the same way. We then drew a smooth curve through those points, and took out the grid.

Given a graph like that we have shown, you can extract the data by essentially working backwards. To find the solubility of KNO_3 at 60°C, draw the dashed line up from 60°C. The solubility is given by the point at which that line intersects the KNO_3 graph line. Draw a horizontal line from that intersection to the vertical axis and read off the solubility of KNO_3, which you can see is about 105 g/100 g H_2O. If you wanted to find the amount of water you would need to dissolve 21 g KNO_3 at 100°C, you would draw the vertical line at 100°C up to the KNO_3 line; then draw the horizontal line at the intersection with the KNO_3 curve, and you find that at 100°C, about 240 g KNO_3 will dissolve in 100 g H_2O. Then a simple conversion factor calculation would give you the amount of water needed to dissolve 21 g KNO_3. Clearly, there is a lot of information in Figure V.1 if you interpret it completely. Its meaning is almost, but not quite, obvious.

Making a Graph

Now let us construct a graph from some experimental data. Given the grid in Figure V.2, in Experiment 7 we measure the pressure of a sample of air as a function of temperature. One of our students obtained the following set of data:

Pressure in mm Hg	674	739	784	821
Temperature in °C	0.0	24.5	42.0	60.1

To make the graph we need to decide what goes where. Since the pressure is dependent on the temperature, we put the pressure on the vertical axis, and the temperature on the horizontal axis, labeling those axes. The temperature goes from 0°C to 60°C; we want the graph to fill the grid, not just be in one corner, so we select a temperature interval between grid lines that will fill the grid. Since we have about 30 vertical grid lines available, and need to cover a 60°C interval, we put the 10° hashmarks at intervals of five grid lines, with 2° between grid lines. Since there are about 20 horizontal grid lines, and we need to cover a pressure change of about 150 mm Hg, we make the lowest pressure 650 mm Hg and the interval between grid lines equal to 10 mm Hg. We put in the hashmarks at 50-mm intervals. Having laid out the graph, we insert each of the data points, using x's or dots. The points fall on a nearly straight line, so we draw such a line, minimizing as best we can the sum of the distances from the data points to the line. If we wish, we can find the equation for that line, which, since it is straight, is of the form $y = mx + b$ where m is the slope of the line and b is the value of y when x equals zero. Using Cricket Graph, our student found that the equation was $y = 2.466x + 676.4$. That equation can be used to find another useful piece of information about the properties of gases, but we need not go into that here.

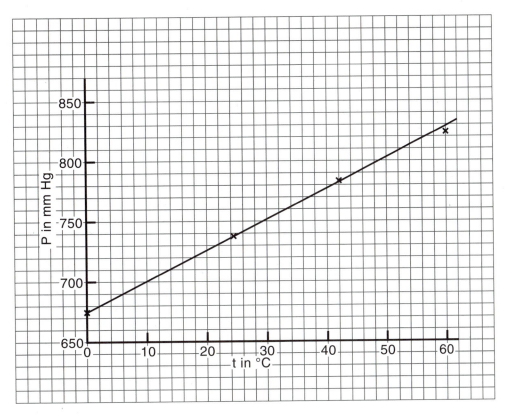

FIGURE V.2

The key to successful construction of graphs is to properly select the variables and assign them to the two axes. Label those axes. Then, given the changes in the variables, select grid intervals that will use up the area of the grid. Put in the hashmarks for each variable at appropriate intervals and label them with the values of the variable. Take the data points and enter them on the grid. If the line is theoretically straight, draw a straight line through the data points in such a way as to minimize the sum of the distances from the points to the line.

Many natural laws are not as simple as the one relating gas pressure to temperature. However, by a few rather easy tricks, involving choice of variables, one can obtain data points that will lie on a straight line. For example, in Experiment 14, we find that the vapor pressure of a liquid depends on temperature according to the following equation:

$$\log_{10} VP = \frac{-\Delta H_{vap}}{2.30\ RT} + C$$

where $\log_{10} VP$ is the logarithm of the vapor pressure to the base 10, T is the Kelvin temperature, and ΔH_{vap} and R are constant and equal to the heat of vaporization and the gas constant, respectively. The equation looks formidable and is. However, if we let x equal $\log_{10} VP$ and $y = 1/T$, then the equation takes the form $y = mx + b$, a straight line with a slope m equal to $-\Delta H_{vap}/2.30\ R$. By using those variables, we can find the slope of the straight line and determine ΔH_{vap} for the liquid. It is very possible that the equation was initially found by an imaginative scientist who measured the vapor pressure against temperature and tried plotting the data against different variables, seeking a straight line dependence. From such efforts many great discoveries have been made.

SUGGESTED LOCKER EQUIPMENT

2 beakers, 30 or 50 mL
2 beakers, 100 mL
2 beakers, 250 mL
2 beakers, 400 mL
1 beaker, 600 cc
2 Erlenmeyer flasks, 25 or 50 mL
2 Erlenmeyer flasks, 125 mL
2 Erlenmeyer flasks, 250 mL
1 graduated cylinder, 10 mL
1 graduated cylinder, 25 or 50 mL
1 funnel, long or short stem
1 thermometer
2 watch glasses, 75, 90, or 100 mm
1 crucible and cover, size #0
1 evaporating dish, small

2 medicine droppers
2 regular test tubes, 18×150 mm
8 small test tubes, 13×100 mm
4 micro test tubes, 10×75 mm
1 test tube brush
1 file
1 spatula
1 test tube holder, wire
1 test tube rack
1 tongs
1 sponge
1 towel
1 plastic wash bottle
1 casserole, small

Appendix VII

SUGGESTIONS FOR EXTENSION OF THE EXPERIMENTS TO "REAL WORLD PROBLEMS"

Our main purpose in writing this laboratory manual was to illustrate as best we could some of the basic chemical principles with reasonably simple experiments.

Several of the experiments can be modified slightly and used to determine the composition or properties of various chemicals we may encounter in our daily lives, including, for example food products, medicinals and minerals.

During the course, your laboratory instructor may assign one or more of the suggested problems as an extension or variation of one of the experiments in the manual. Or, you may be allowed to try to do one of the problems as a special project.

In the experiments in the manual we furnish very specific procedures. If you work on a suggested problem, you will have to modify the procedure to some degree to apply it to that problem. You will have to at least consider the size of the sample to work with, but there may be other factors involving preliminary treatment of the sample, such as grinding, filtering off extraneous material, or setting the pH that need to be dealt with. In most cases the changes necessary are small, but it is possible that in some of the problems you will run into unexpected difficulties. The practicing chemist encounters similar difficulties in his or her work. The solutions are not always obvious, but when found, may offer leads to new areas to study. With a little practice, you should be able to design your experiments and carry them out successfully. Before beginning work, however, you should check with your instructor to see that your approach makes sense and does not involve any dangerous reactions or reagents.

Experiment 6. Find the per cent Cl^- in table salt, a bouillon cube, or canned chicken or French onion soup. Assuming that all of the Cl^- comes from NaCl, calculate the per cent NaCl in the sample.

As a class project, measure the temperature of a series of equilibrium mixtures of ice-NaCl solutions, up to the point where the solution is saturated with NaCl. The temperature at that point is about $-21°C$. Find the % NaCl in each solution. Plot the temperature, which is the freezing-point of the mixture vs. % NaCl.

Experiment 9. Find the per cent Al in an aluminum soda or beer can.

Experiment 13. Measure the heat of neutralization of household vinegar in its reaction with NaOH solution. Given the percent Acetic Acid from the next suggested problem, find the molar heat of neutralization of Acetic Acid.

Experiment 22. Find the mass per cent Acetic Acid in household vinegar.

Find the neutralization capacity of an antacid tablet, such as Tums, Rolaids, or Alka Seltzer.

Experiment 23. Study the buffering properties of Vitamin C, ascorbic acid (MM = 176 g), using the optional part of the experiment. Find K_a for Vitamin C, and the pH range over which it might be useful as a buffer.

Red cabbage extract and grape juice may be useful as natural acid-base indicators. Find the color of the acidic and basic forms, the value of K_a, and the pH range of one of these materials.

Experiment 26. Find the per cent $CaCO_3$ in a Tums tablet, assuming that all of the calcium is in the form of $CaCO_3$.

Find the per cent $CaCO_3$ in a sample of limestone.

Experiment 34. Use qualitative analysis to establish the presence of copper ion in a mineral supplement tablet. Then estimate the amount of copper present using the procedure in Experiment 31.